W9-AWI-265

STATA
STRUCTURAL EQUATION MODELING
REFERENCE MANUAL
RELEASE 12

A Stata Press Publication
StataCorp LP
College Station, Texas

Copyright © 1985–2011 StataCorp LP
All rights reserved
Version 12

Published by Stata Press, 4905 Lakeway Drive, College Station, Texas 77845
Typeset in TeX
Printed in the United States of America
10 9 8 7 6 5 4 3 2 1

ISBN-10: 1-59718-097-1
ISBN-13: 978-1-59718-097-9

This manual is protected by copyright. All rights are reserved. No part of this manual may be reproduced, stored in a retrieval system, or transcribed, in any form or by any means—electronic, mechanical, photocopy, recording, or otherwise—without the prior written permission of StataCorp LP unless permitted subject to the terms and conditions of a license granted to you by StataCorp LP to use the software and documentation. No license, express or implied, by estoppel or otherwise, to any intellectual property rights is granted by this document.

StataCorp provides this manual "as is" without warranty of any kind, either expressed or implied, including, but not limited to, the implied warranties of merchantability and fitness for a particular purpose. StataCorp may make improvements and/or changes in the product(s) and the program(s) described in this manual at any time and without notice.

The software described in this manual is furnished under a license agreement or nondisclosure agreement. The software may be copied only in accordance with the terms of the agreement. It is against the law to copy the software onto DVD, CD, disk, diskette, tape, or any other medium for any purpose other than backup or archival purposes.

The automobile dataset appearing on the accompanying media is Copyright © 1979 by Consumers Union of U.S., Inc., Yonkers, NY 10703-1057 and is reproduced by permission from CONSUMER REPORTS, April 1979.

Stata, STATA Stata Press, Mata, MATA and NetCourse are registered trademarks of StataCorp LP.

Stata and Stata Press are registered trademarks with the World Intellectual Property Organization of the United Nations.

NetCourseNow is a trademark of StataCorp LP.

Other brand and product names are registered trademarks or trademarks of their respective companies.

For copyright information about the software, type `help copyright` within Stata.

The suggested citation for this software is

StataCorp. 2011. *Stata: Release 12*. Statistical Software. College Station, TX: StataCorp LP.

Table of contents

Cross-referencing the documentation

When reading this manual, you will find references to other Stata manuals. For example,

[U] **26 Overview of Stata estimation commands**
[XT] **xtabond**
[D] **reshape**

The first example is a reference to chapter 26, *Overview of Stata estimation commands*, in the *User's Guide*; the second is a reference to the `xtabond` entry in the *Longitudinal-Data/Panel-Data Reference Manual*; and the third is a reference to the `reshape` entry in the *Data-Management Reference Manual*.

All the manuals in the Stata Documentation have a shorthand notation:

[GSM] *Getting Started with Stata for Mac*
[GSU] *Getting Started with Stata for Unix*
[GSW] *Getting Started with Stata for Windows*
[U] *Stata User's Guide*
[R] *Stata Base Reference Manual*
[D] *Stata Data-Management Reference Manual*
[G] *Stata Graphics Reference Manual*
[XT] *Stata Longitudinal-Data/Panel-Data Reference Manual*
[MI] *Stata Multiple-Imputation Reference Manual*
[MV] *Stata Multivariate Statistics Reference Manual*
[P] *Stata Programming Reference Manual*
[SEM] *Stata Structural Equation Modeling Reference Manual*
[SVY] *Stata Survey Data Reference Manual*
[ST] *Stata Survival Analysis and Epidemiological Tables Reference Manual*
[TS] *Stata Time-Series Reference Manual*
[I] *Stata Quick Reference and Index*

[M] *Mata Reference Manual*

Detailed information about each of these manuals may be found online at

http://www.stata-press.com/manuals/

Title

intro 1 — Introduction

Description

SEM stands for structural equation modeling. SEM is

1. A notation for specifying structural equation models.
2. A way of thinking about structural equation models.
3. Methods for estimating the parameters of structural equation models.

Stata's `sem` command implements linear structural equation models.

Remarks

Do you know what SEM is? It encompasses a broad array of models from linear regression to measurement models to simultaneous equations, including along the way confirmatory factor analysis (CFA), correlated uniqueness models, latent growth models, and multiple indicators and multiple causes (MIMIC).

SEM is not just an estimation method for a particular model in the way that Stata's `regress` and `probit` commands are, or even in the way that `stcox` and `xtmixed` are. SEM is a way of thinking, a way of writing, and a way of estimating.

If you read the introductory manual pages in the front of this manual—[SEM] **intro 1**, [SEM] **intro 2**, and so on—we will do our best to familiarize you with SEM and our implementation, `sem`.

Beginning with [SEM] **intro 2**, entitled *Learning the language: Path diagrams and command language*, you will learn that

1. A particular SEM is usually described using a path diagram.
2. Stata's `sem` command allows you to use path diagrams to input models.
3. Stata's `sem` command allows you to use a command language to input models. This command language is similar to path diagrams.

Then in [SEM] **intro 3**, entitled *Substantive concepts*, you will learn that

4. Stata's `sem` provides four different estimation methods; you need to specify the method appropriate for the assumptions you are willing to make.
5. There are four types of variables in SEM: observed versus latent and endogenous versus exogenous. To this, `sem` adds a fifth: error variables, which are latent exogenous variables with a fixed unit path coefficient and associated with a single endogenous variable. Error variables are denoted with an `e.` prefix, so if `x1` is an endogenous variable, then `e.x1` is the associated error variable.
6. It is easy to specify path constraints in SEM—you just draw them on the diagram. It is similarly easy using `sem`'s GUI. It is similarly easy using `sem`'s command language.
7. Determining whether an SEM is identified can be difficult. We show you how to let the software check for you.
8. Identification also includes normalization constraints. `sem` applies normalization constraints automatically, but you can control that if you wish. Sometimes you might even need to control it.

9. If you think identification is a bear, wait until you hear about starting values. We figure 5% to 15% of complicated models will cause you difficulty. We show you what to do and it is not difficult.

Then in [SEM] **intro 4**, entitled *Tour of models*,

10. We take you on a whirlwind tour of some of the models that SEM and `sem` can fit. This is a fun and useful section because we give you a real overview and do not get lost in the details.

Then in [SEM] **intro 5**, entitled *Comparing groups*,

11. We show you a highlight of `sem`: its ability to take an SEM model and data consisting of groups—sexes, age categories, and the like—and fit the model in an interacted way that makes it easy for you to test whether and how the groups differ.

In [SEM] **intro 6**, entitled *Postestimation tests and predictions*,

12. We show you how to redisplay results and to obtain standardized results.

13. We show you how to obtain goodness-of-fit statistics.

14. We show you how to perform hypothesis tests, including tests for omitted paths, tests for relaxing constraints, and tests for model simplification.

15. We show you how to display other results, statistics, and tests.

16. We show you how to obtain predicted values, including predicted factor scores.

17. We show you how to access saved results.

In [SEM] **intro 7**, entitled *Robust and clustered standard errors*,

18. We mention that `sem` optionally provides robust standard errors and that `sem` provides clustered standard errors, which relaxes the assumption of independence of observations to independence within clusters of observations.

In [SEM] **intro 8**, entitled *Standard errors, the full story*,

19. We provide lots of technical detail expanding on item 18.

In [SEM] **intro 9**, entitled *Fitting models using survey data*,

20. We explain how `sem` can be used with Stata's `svy:` prefix to obtain results adjusted for complex survey designs, including clustered sampling and stratification.

Finally, in [SEM] **intro 10**, entitled *Fitting models using summary statistics data*,

21. We will show you how to use `sem` with summary statistics data such as the correlation or covariance matrix rather than the raw data. Many sources, and especially textbooks, publish data in summary statistics form.

In the meantime,

22. There are many examples that we have collected for you in [SEM] **example 1**, [SEM] **example 2**, and so on. It is entertaining and informative simply to read the examples in order.

23. There is an alphabetical glossary in [SEM] **Glossary**, located at the end of the manual.

If you prefer, you can skip all this introductory material and go for the details. For the full experience, go directly to [SEM] **sem**. You will have no idea what we are talking about—we promise.

The technical sections, in logical order, are

Estimation
- [SEM] **sem**
- [SEM] **sem path notation**
- [SEM] **GUI**
- [SEM] **sem model description options**
- [SEM] **sem group options**
- [SEM] **sem ssd options**
- [SEM] **sem estimation options**
- [SEM] **sem reporting options**
- [SEM] **sem syntax options**
- [SEM] **sem option noxconditional**
- [SEM] **sem option select()**
- [SEM] **sem option covstructure()**
- [SEM] **sem option method()**
- [SEM] **sem option reliability()**
- [SEM] **sem option from()**
- [SEM] **sem option constraints()**
- [SEM] **ssd**

Postestimation, summary of
- [SEM] **sem postestimation**

Reporting results
- [R] **estat**
- [SEM] **estat teffects**
- [SEM] **estat residuals**
- [SEM] **estat framework**

Goodness-of-fit tests
- [SEM] **estat gof**
- [SEM] **estat eqgof**
- [SEM] **estat ggof**
- [R] **estat**

Hypotheses tests
- [SEM] **estat mindices**
- [SEM] **estat eqtest**
- [SEM] **estat scoretests**
- [SEM] **estat ginvariant**
- [SEM] **estat stable**
- [SEM] **test**
- [SEM] **lrtest**
- [SEM] **testnl**
- [SEM] **estat stdize**

Linear and nonlinear combinations of results
- [SEM] **lincom**
- [SEM] **nlcom**

Predicted values
- [SEM] **predict**

These sections are technical, but mostly in the computer sense of the word. We suggest that when you read the technical sections, you immediately skip to *Remarks*. If you read the introductory sections, you will already know how to use the commands, so there is little reason to confuse yourself with syntax diagrams that are more precise than they are enlightening. However, the syntax diagrams do serve as useful reminders.

Also see

[SEM] **intro 2** — Learning the language: Path diagrams and command language

[SEM] **example 1** — Single-factor measurement model

Title

intro 2 — Learning the language: Path diagrams and command language

Description

Individual structural equation models are usually described using path diagrams. Path diagrams are described here.

Path diagrams can be used in `sem`'s GUI as the input to describe the model to be fit. Path diagrams differ a little from author to author, and `sem`'s path diagrams differ a little, too. We omit drawing the variances and covariances between observed exogenous variables by default.

`sem` also provides a command-language interface. This interface is similar to path diagrams but is typable.

Remarks

Remarks are presented under the following headings:

Using path diagrams to specify the model

In SEM, models are often illustrated in a path diagram, such as

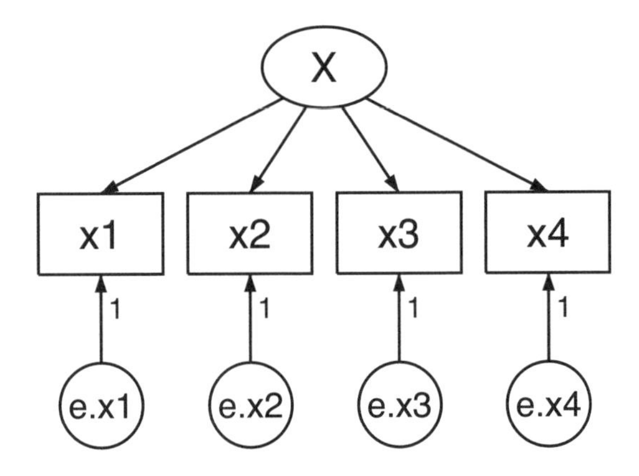

This diagram is composed of the following:

1. Boxes and circles with variable names written inside them.

    a. Boxes contain variables that are observed in the data.

    b. Circles contain variables that are unobserved, known as latent variables.

2. Arrows, called paths, that connect some of the boxes and circles.

    a. When a path points from one variable to another, it means that the first variable affects the second.

b. More precisely, if $s \rightarrow d$, it means to add $\beta_k s$ to the linear equation for d. β_k is called the path coefficient.

c. Sometimes small numbers are written along the arrow connecting two variables. This means that β_k is constrained to be the value specified.

d. When no number is written along the arrow, the corresponding coefficient is to be estimated from the data. Sometimes symbols are written along the path arrow to emphasize this and sometimes not.

e. The same path diagram used to describe the model can be used to display the results of estimation. In that case, estimated coefficients appear along the paths.

3. There are other elements that may appear on the diagram to indicate variances and between-variable correlations. We will get to them later.

Thus the above figure corresponds to the equations

$$x_1 = \alpha_1 + \beta_1 X + e.x_1$$

$$x_2 = \alpha_2 + \beta_2 X + e.x_2$$

$$x_3 = \alpha_3 + \beta_3 X + e.x_3$$

$$x_4 = \alpha_4 + \beta_4 X + e.x_4$$

There's a third way of writing this model, namely,

```
(x1<-X) (x2<-X) (x3<-X) (x4<-X)
```

This is the way we would write the model if we wanted to use `sem`'s command syntax rather than drawing the model in `sem`'s GUI. The full command we would type would be

```
. sem (x1<-X) (x2<-X) (x3<-X) (x4<-X)
```

We will get to that later.

However we write this model, what is it? It is a *measurement model*, a term loaded with meaning for some researchers. X might be mathematical ability. x_1, x_2, x_3, and x_4 might be scores from tests designed to measure mathematical ability. x_1 might be the score based on your answers to a series of questions after reading this section.

The model we have just drawn, or written in mathematical notation, or written in Stata command notation, can be interpreted in other ways, too. Look at this diagram:

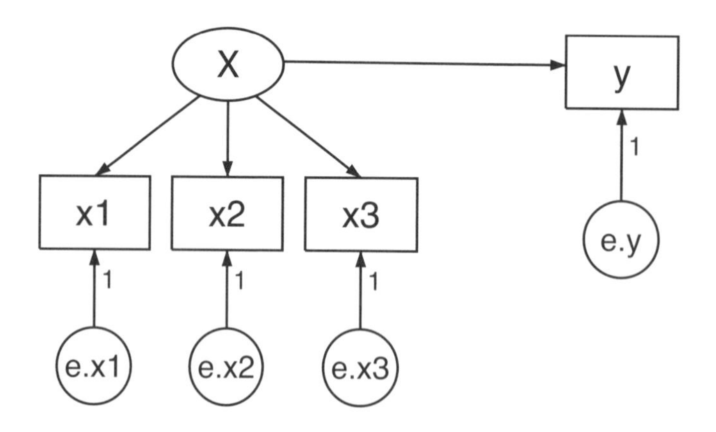

Despite appearances, this diagram is identical to the previous diagram except that we have renamed x_4 to be y. The fact that we changed a name obviously does not matter substantively. The fact that we have rearranged the boxes in the diagram is irrelevant, too; paths connect the same variables in the same directions. The equations for the above diagrams are the same as the previous equations with the substitution of y for x_4:

$$x_1 = \alpha_1 + X\beta_1 + e.x_1$$

$$x_2 = \alpha_2 + X\beta_2 + e.x_2$$

$$x_3 = \alpha_3 + X\beta_3 + e.x_3$$

$$y = \alpha_4 + X\beta_4 + e.y$$

The Stata command notation changes similarly:

```
(x1<-X) (x2<-X) (x3<-X) (y<-X)
```

Many people looking at the model written this way might decide that it is not a measurement model but a measurement error model. y depends on X but we do not observe X. We do observe x_1, x_2, and x_3, each a measurement of X, but with error. Our interest is in knowing β_4, the effect of true X on y.

A few others might disagree and instead see a model for interrater agreement. Obviously, we have four raters who each make a judgment, and we want to know how well the judgment process works and how well each of these raters performs.

Specifying correlation

One of the key features of SEM is the ease with which you can allow correlation between latent variables to adjust for the reality of the situation. In the measurement model in the previous section, we drew the following path diagram:

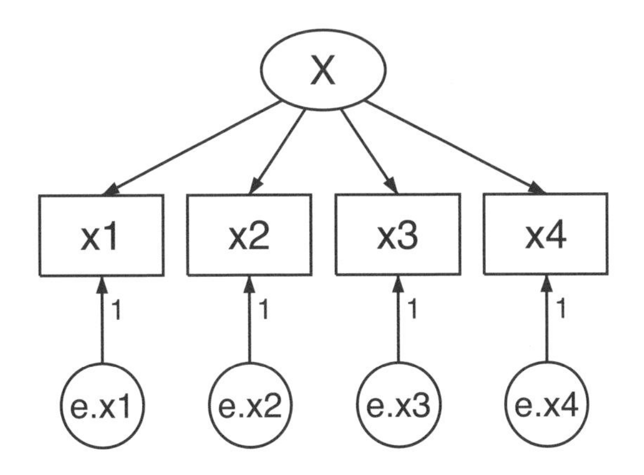

which corresponded to the equations

$$x_1 = \alpha_1 + X\beta_1 + e.x_1$$

$$x_2 = \alpha_2 + X\beta_2 + e.x_2$$

$$x_3 = \alpha_3 + X\beta_3 + e.x_3$$

$$x_4 = \alpha_4 + X\beta_4 + e.x_4$$

$$(X, x_1, x_2, x_3, x_4, e.x_1, e.x_2, e.x_3, e.x_4) \sim \text{i.i.d. with mean } \mu \text{ and variance } \boldsymbol{\Sigma}$$

where i.i.d. means that observations are independent and identically distributed.

We must appreciate that μ and $\boldsymbol{\Sigma}$ are estimated, just as are α_1, β_1, ..., α_4, β_4. Some of the elements of $\boldsymbol{\Sigma}$, however, are constrained to be zero; which elements are constrained is determined by how we specify the model. In the above diagram, we drew the model in such a way that we assumed that error variables were uncorrelated with each other. We could have drawn the diagram differently. If we wish to allow for a correlation between $e.x_2$ and $e.x_3$, we add a curved path between the variables:

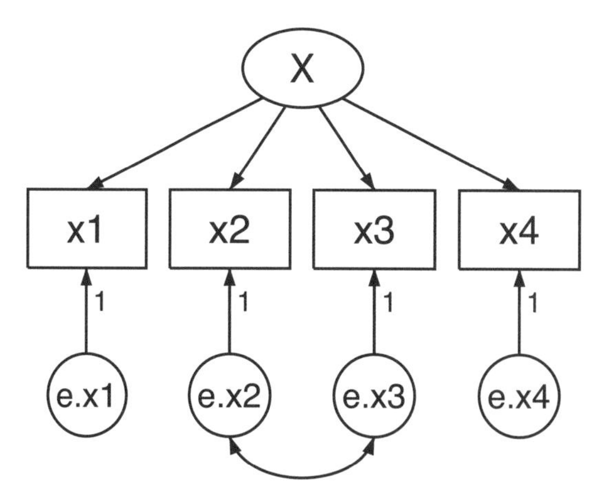

The curved path states that there is a correlation to be estimated between the variables it connects. The absence of a curved path—say, between $e.x_1$ and $e.x_4$—means the variables are constrained to be uncorrelated. That is not to say that x_1 and x_4 are uncorrelated. Obviously, they are correlated because both are functions of the same X. Their corresponding error variables, however, are uncorrelated. The equations for this model, in their full detail, are

$$x_1 = \alpha_1 + X\beta_1 + e.x_1$$

$$x_2 = \alpha_2 + X\beta_2 + e.x_2$$

$$x_3 = \alpha_3 + X\beta_3 + e.x_3$$

$$x_4 = \alpha_4 + X\beta_4 + e.x_4$$

$$(X, x_1, x_2, x_3, x_4, e.x_1, e.x_2, e.x_3, e.x_4) \sim \text{i.i.d. with mean } \mu \text{ and variance } \boldsymbol{\Sigma}$$

$$
\begin{aligned}
\boldsymbol{\Sigma} \text{ is constrained such that} \\
\sigma_{e.x_1,e.x_2} = \sigma_{e.x_2,e.x_1} = 0 \\
\sigma_{e.x_1,e.x_3} = \sigma_{e.x_3,e.x_1} = 0 \\
\sigma_{e.x_1,e.x_4} = \sigma_{e.x_4,e.x_1} = 0 \\
\sigma_{e.x_2,e.x_4} = \sigma_{e.x_4,e.x_2} = 0 \\
\sigma_{e.x_3,e.x_4} = \sigma_{e.x_4,e.x_3} = 0 \\
\\
\sigma_{X,e.x_1} = \sigma_{e.x_1,X} = 0 \\
\sigma_{X,e.x_2} = \sigma_{e.x_2,X} = 0 \\
\sigma_{X,e.x_3} = \sigma_{e.x_3,X} = 0 \\
\sigma_{X,e.x_4} = \sigma_{e.x_4,X} = 0 \\
\mu \text{ is constrained such that} \\
\mu_X = 0 \\
\mu_{e.x_1} = 0 \\
\mu_{e.x_2} = 0 \\
\mu_{e.x_3} = 0 \\
\mu_{e.x_4} = 0 \\
\text{The parameters to be estimated are} \\
\alpha_1, \beta_1, \ldots, \alpha_4, \beta_4, \\
\mu \\
\sigma
\end{aligned}
$$

What is missing from the above list of constraints and which was included in the previous list, the list we never showed you, is

$$\sigma_{e.x_2,e.x_3} = \sigma_{e.x_3,e.x_2} = 0$$

The line is missing because we drew a curved path between $e.x_2$ and $e.x_3$. The line was previously included because we had not drawn the curved path.

There are lots of other curved arrows we could have drawn. By not drawing them, we are asserting that the corresponding covariance is zero.

Some authors would draw the above model as

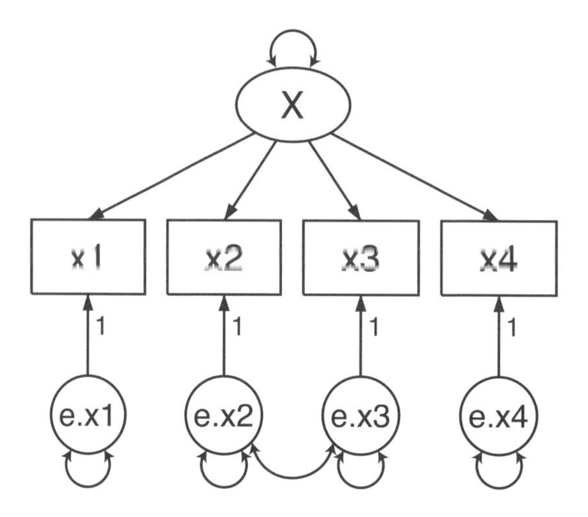

Notice the curved paths from the latent variables back to themselves. A curved path from a variable to itself indicates a variance. All covariances could be shown, including between latent variables and between observed exogenous variables.

When we draw diagrams, however, we will assume variance paths and omit drawing them, and we will similarly assume and omit drawing covariances between observed exogenous variables. The GUI has an option concerning the latter. Like everyone else, we will not assume correlations between latent variables unless they are shown.

In sem's command-language notation, curved paths between variables are indicated via an option:

```
(x1<-X) (x2<-X) (x3<-X) (x4<-X), cov(e.x2*e.x3)
```

In this notation, the cov() option does not appear to indicate that covariances are constrained to zero except for covariances between exogenous variables. Exogenous variables are assumed to be correlated, observed exogenous variables are assumed to be correlated, latent exogenous variables are assumed to be correlated, and observed and latent variables are assumed to be correlated.

Using the command language to specify the model

You can describe your model to sem by using path diagrams with sem's GUI interface, or you can describe your model by using sem's command language. Here are the trade-offs:

1. If you use path diagrams, you can see the results of your estimation as path diagrams or standard computer output.
2. If you use the command language, only standard computer output is available.
3. Typing models in command language is usually quicker than drawing them in the GUI.
4. You can type models in the command language and store them in do-files. By doing so, you can more easily correct the errors you make.

Translating from path diagrams to command language is easy.

1. Path diagrams have squares and circles to distinguish observed from latent variables.

   In the command language, variables are assumed to be observed if they are typed in lowercase and are assumed to be latent if the first letter is capitalized. Variable `educ` is observed. Variable `Knowledge` or `KNOWLEDGE` is latent.

   If the observed variables in your dataset have uppercase names, type `rename _all, lower` to covert them to lowercase; see [D] **rename group**.

2. When typing path diagrams in the command language, remember the `///` continuation line indicator. You may type

```
. sem (x1<-X) (x2<-X) (x3<-X) (y<-X)
```

   or you may type

```
. sem (x1<-X) (x2<-X)      ///
      (x3<-X) (y<-X)
```

   or you may type

```
. sem (x1<-X)              ///
      (x2<-X)              ///
      (x3<-X)              ///
      (y<-X)
```

3. In path diagrams, you draw arrows connecting variables to indicate paths. In the command language, you type variable names and arrows. Arrows may be typed in either direction. The following mean the same thing:

```
(x1 <- X)
(X -> x1)
```

4. In the command language, you may type multiple variables on either side of the arrow:

```
(X -> x1 x2 x3 x4)
```

   The above means the same as

```
(X -> x1) (X -> x2) (X -> x3) (X -> x4)
```

   which means the same as

```
(x1 <- X) (x2 <- X) (x3 <- X) (x4 <- X)
```

   which means the same as

```
(x1 x2 x3 x4 <- X)
```

   In a more complicated measurement model, we might have

```
(X Y -> x1 x2 x3) (X -> x4 x5) (Y -> x6 x7)
```

   The above means the same as

```
(X -> x1 x2 x3 x4 x5)  ///
(Y -> x1 x2 x3 x6 x7)
```

which means

```
(X->x1) (X->x2) (X->x3) (X->x4) (X->x5)     ///
(Y->x1) (Y->x2) (Y->x3) (Y->x6) (Y->x7)
```

5. In path diagrams, you are required to show the error variables. In the command language, you may omit the error variables. `sem` knows that each endogenous variable needs an error variable. You can type

```
(x1 <- X) (x2 <- X) (x3 <- X) (x4 <- X)
```

and that means the same thing as

```
(x1 <- X e.x1)   ///
(x2 <- X e.x2)   ///
(x3 <- X e.x3)   ///
(x4 <- X e.x4)
```

except that we have lost the small numeric ones we had next to the arrows from `e.x1` to `x1`, `e.x2` to `x2`, and so on. To constrain the path coefficient, you type

```
(x1 <- X e.x1@1)    ///
(x2 <- X e.x2@1)    ///
(x3 <- X e.x3@1)    ///
(x4 <- X e.x4@1)
```

It is easier to simply type

```
(x1 <- X) (x2 <- X) (x3 <- X) (x4 <- X)
```

but now you know that if you wanted to constrain the path coefficient `x2<-X` to be 2, you could type

```
(x1 <- X) (x2 <- X@2) (x3 <- X) (x4 <- X)
```

If you wanted to constrain the path coefficients `x2<-X` and `x3<-X` to be equal, you could type

```
(x1 <- X) (x2 <- X@b) (x3 <- X@b) (x4 <- X)
```

We have not covered symbolic constraints with path diagrams yet, but typing symbolic names along the paths in either the GUI or the command language is how you constrain coefficients to be equal.

6. Curved paths are specified using the `cov()` option after you have specified your model:

```
(x1 x2 x3 x4<-X), cov(e.x2*e.x3)
```

If you wanted to allow for correlation of `e.x2*e.x3` and `ex3*e.x4`, you can specify that in a single `cov()` option,

```
(x1 x2 x3 x4<-X), cov(e.x2*e.x3 e.x3*ex4)
```

or in separate `cov()` options:

```
(x1 x2 x3 x4<-X), cov(e.x2*e.x3) cov(e.x3*ex4)
```

Also see

[SEM] **intro 1** — Introduction

[SEM] **intro 3** — Substantive concepts

[SEM] **GUI** — Graphical user interface

[SEM] **sem path notation** — Command syntax for path diagrams

Title

intro 3 — Substantive concepts

Description

The SEM way of describing models is deceptively simple. It is deceptive because the machinery underlying SEM is sophisticated, complex, and sometimes, well, at least temperamental if not mercurial, and it can be temperamental both in substantive statistical ways and in practical computer ways.

Professional researchers need to understand these issues.

Remarks

Remarks are presented under the following headings:

Assumptions and choice of estimation method

Let's return to our measurement model:

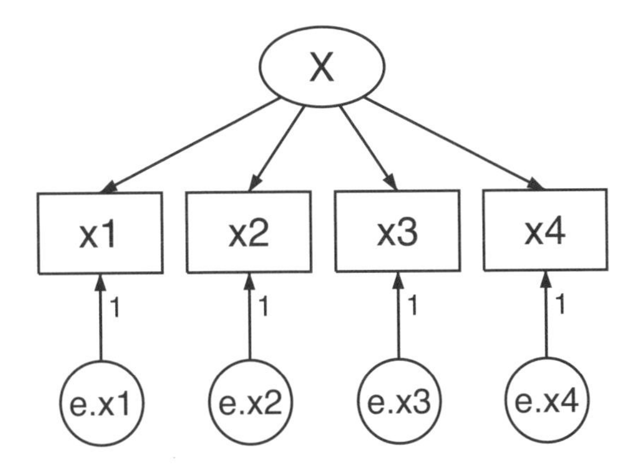

We will show you other, more interesting models later. Right now, this model is complicated enough for our purposes.

What is being estimated and the assumptions we are making

The corresponding mathematical equations for this model are

$$x_1 = \alpha_1 + X\beta_1 + e.x_1$$

$$x_2 = \alpha_2 + X\beta_2 + e.x_2$$

$$x_3 = \alpha_3 + X\beta_3 + e.x_3$$

$$x_4 = \alpha_4 + X\beta_4 + e.x_4$$

The mathematical equations we have written above are an exact translation of what we have drawn in the diagram. To finish off the mathematical description of, say, the measurement model in which the observed variables were named $x_1, \ldots, x_4$, it is usual in SEM modeling to add

$$(X, x_1, x_2, x_3, x_4, e.x_1, e.x_2, e.x_3, e.x_4) \sim N(\mu, \mathbf{\Sigma})$$

which is to say that we assume that all variables, both observed and latent, follow a multivariate normal distribution. Observations are assumed independent. The entire model becomes

$$x_1 = \alpha_1 + X\beta_1 + e.x_1$$

$$x_2 = \alpha_2 + X\beta_2 + e.x_2$$

$$x_3 = \alpha_3 + X\beta_3 + e.x_3$$

$$x_4 = \alpha_4 + X\beta_4 + e.x_4$$

$$(X, x_1, x_2, x_3, x_4, e.x_1, e.x_2, e.x_3, e.x_4) \sim N(\mu, \mathbf{\Sigma})$$

and to be estimated is

$$\theta = \{\alpha, \beta, \mu, \Sigma\}$$

It can be proven that if we assume the joint normality of the variables, consistent and asymptotically normal estimates of θ can be obtained.

Joint normality can be too restrictive

For some researchers, the assumption of joint normality is not overly restrictive. For others, the assumption is too restrictive to encompass the data they analyze. It turns out that the normality assumption is usually unnecessary and one can substitute conditional normality in its place.

In the measurement model, the assumption of normality would mean, for instance, that observed variable `x1` is distributed normal. Because `x1` is observed, we could look at its values and judge the reasonableness of that assumption. Based on what we see, we might find the assumption difficult or impossible to postulate.

Conditional normality might be sufficient

Conditional normality means that we instead take the values of the observed variables as given, which in this model would mean the values of x_1, x_2, x_3, and x_4. Assuming the model specification is correct and $e.x_1$, $e.x_2$, $e.x_3$, and $e.x_4$ are not dependent on X, there are only three ways these variables cannot be normally distributed:

1. $e.x_1$, $e.x_2$, $e.x_3$, and $e.x_4$ are not distributed normal, or
2. X is not distributed normal, or
3. both items 1 and 2.

Focus on the second possibility for the moment. If we condition on X, we are left with assuming

$$(e.x_1, e.x_2, e.x_3, e.x_4) \sim N(\mu_r, \Sigma_r)$$

or, equivalently,

$$(x_1, x_2, x_3, x_4)|X \sim N(\mu_x, \Sigma_x)$$

So what happens if we substitute this less restrictive assumption for the usual assumption? The mathematics is difficult. Simulations, however, strongly suggest that the only problem is that the standard error of the estimate of the variance of X will be poor, or at least it will be if we use the maximum likelihood (ML) estimator to obtain solutions for the parameters of the model.

How the estimation methods respond to conditional normality

Strictly speaking, the assumptions that one must make to establish the consistency of the estimates and their asymptotic normality is determined by the method used to estimate them. Stata's `sem` command provides four methods: ML, QML, ADF, and MLMV. We need to discuss each separately.

1. ML stands for maximum likelihood and is the method that `sem` uses by default. It is easy to derive that results are consistent and asymptotically normal if one assumes joint normality of all the variables, but usually one can derive most of the desired properties from conditional normality. The result will be that across all models, estimated parameters will be consistent and asymptotically normal except for parameters reflecting paths or covariances between conditioning variables. You have to work out for your model what the conditioning variables are. In this case, we worked out together that there is only one conditioning variable, namely, X. Thus the only possible "between conditioning variables" is X with itself, and thus the estimated variance of X and its standard error are suspect. Simulations reveal that the variance estimate tends to be consistent despite lack of a theoretical guarantee. Unfortunately, but not unexpectedly, the standard error turned out poorly in the simulations.

2. QML, quasimaximum likelihood, uses ML to fit the model but relaxes the normality assumptions when estimating the standard errors. Thus concerning the parameter estimates, everything just said about ML applies to QML. Concerning standard errors, theoretically we expect consistent standard errors, and practically that is what we observe in our simulations. Be aware that QML does not fix the theoretical concerns about the estimated parameter for the variance of X; it just turns out that for this model, the parameter is well estimated. We suspect that parameters corresponding to paths between conditioning variables are okay in other models as well, but we cannot verify that for you across all models. `sem` uses QML when you specify both the `method(ml)` and `vce(robust)` options.

3. ADF estimates are produced when you specify `sem`'s `method(adf)` option. ADF stands for asymptotic distribution free, and it makes no assumption of joint normality or even symmetry. ADF is a form of weighted least squares (WLS). ADF is also a generalized method of moments (GMM) estimator. You are on firm theoretical ground treating the observed variables as given. In simulations of the measurement model, ADF produces excellent results, even for the standard error of the variance of X. Be aware, however, that ADF is less efficient than ML when ML's assumptions hold, whatever those minimal assumptions are. On the other hand, under the less restrictive ADF assumptions, the ADF estimator is more efficient than QML, although the QML estimator will still be consistent and have correct coverage.

4. MLMV estimates are produced when you specify `sem`'s `method(mlmv)` option. This has to do with using the information in observations containing missing values, observations that are omitted by the other methods; see [SEM] **example 26**. Missing values are assumed to be missing at random (MAR), which is a term asserting that the missing values are not just scattered completely at random throughout the data, but that if some values are more likely to be missing than others, that can be predicted by the variables in the model. MLMV takes the assumption of joint normality seriously in most cases. If your observed variables do not follow a joint normal distribution, you will be better off using ML, QML, or ADF and simply omitting observations with missing values. The assumption of conditional normality, however, will work well with MLMV when the missing values occur only in endogenous variables.

In the measurement model, the other possible way to eliminate the requirement of joint normality is to eliminate the assumption that $e.x_1$, ..., $e.x_4$ are distributed normally. Let's go back through the methods and consider the ramifications of substituting that assumption.

1. ML parameter estimates will exhibit the same properties described previously. Without at least conditional normality, however, you lose the justification underlying hypothesis testing.

2. QML parameter estimates will exhibit the same properties as ML because they are produced by ML. In QML, standard errors are corrected for nonnormality of the errors.

3. ADF never had any assumption about normality. Parameter estimates will be consistent and the standard errors efficient.

4. MLMV requires full normality.

Thus we will write the measurement model as

$$x_1 = \alpha_1 + X\beta_1 + e.x_1$$

$$x_2 = \alpha_2 + X\beta_2 + e.x_2$$

$$x_3 = \alpha_3 + X\beta_3 + e.x_3$$

$$x_4 = \alpha_4 + X\beta_4 + e.x_4$$

$$(X, x_1, x_2, x_3, x_4, e.x_1, e.x_2, e.x_3, e.x_4) \sim \text{ i.i.d. with mean } \mu \text{ and variance } \boldsymbol{\Sigma}.$$

That is, we are going to be relaxed about assumptions, but you need to consider them and choose the appropriate estimation method given the assumptions that are reasonable for your analysis.

Variable types: Observed, latent, endogenous, exogenous, and error

Structural equation models can contain four different types of variables:

1. observed exogenous
2. observed endogenous
3. latent exogenous
4. latent endogenous

As a software matter, it is useful to think as though there is a fifth type, too:

5. error

Errors are in fact a special case of latent exogenous variables, but there will be good reason to consider them separately.

As a language matter, it is sometimes useful to think of there being yet another type, namely

6. measure or measurement

Measurement variables are a case of item 2, observed endogenous variables.

Let us explain:

Observed.
A variable is observed if it is a variable in your dataset. In this documentation, we often refer to observed variables using `x1`, `x2`, ..., `y1`, `y2`, and so on, but in reality observed variables have names such as `mpg`, `weight`, `testscore`, and so on.

Latent.
A variable is latent if it is not observed. A variable is latent if it is not in your dataset but you wish it were. You wish you had a variable recording the propensity to commit violent crime, or socioeconomic status, or happiness, or true ability, or even income accurately recorded. Sometimes, latent variables are imagined variants of real variables, variables that are somehow better, such as being measured without error. At the other end of the spectrum are latent variables that are not even conceptually measurable.

In this documentation, latent variables usually have names such as L1, L2, F1, ..., but in real life the names are more descriptive such as VerbalAbility, SES, and so on. The sem command assumes variables are latent if the first letter of the name is capitalized, so we will capitalize our latent variable names.

Endogenous.
A variable is endogenous (determined by the system) if any path points to it. (As an aside, endogenous variables are required to have a path from an error variable pointing to them, and this happens automatically in the software.)

Exogenous.
A variable is exogenous (determined outside the system) if paths only originate from it, or equivalently, no path points to it. (Exogenous variables never have a path from an error variable pointing to them.)

With the above definitions, we can now describe the five types of variables:

Observed exogenous.
An observed exogenous variable is a variable in your dataset that is treated as exogenous in your model.

In sem, all exogenous variables—observed and latent—are assumed to be correlated with each other; it is just a matter of estimating what the corresponding covariances are.

sem's GUI has a mode where it will draw curved paths corresponding to the assumption, but usually that clutters the diagram too much. Nonetheless, the covariances are there unless you explicitly draw the path and constrain the covariance to zero.

In the command notation, you can constrain the covariance by using the cov() option.

Observed exogenous variables are assumed to be uncorrelated with all error variables.

Observed endogenous.
An observed endogenous variable is a variable in your dataset that is treated as endogenous in your model.

Covariances between observed endogenous variables and other variables in the model are not estimated directly, but implied by the model. There are no options to specify covariances related to observed endogenous variables in the GUI or in command notation.

Latent exogenous.
A latent exogenous variable is an unobserved variable that is treated as exogenous in your model.

As said above, in sem, all exogenous variables—observed and latent—are assumed to be correlated with each other.

sem's GUI usually shows the assumed covariance paths between latent exogenous variables and other exogenous variables, both latent and observed. If F1 and F2 are latent exogenous variables, to constrain them to be uncorrelated, delete the covariance path or place a zero next to the path.

In the command notation, specify cov(F1*F2@0) to constrain the covariance to be zero.

Latent endogenous.
A latent endogenous variable is an unobserved variable that is treated as endogenous in your model.

Covariances between latent endogenous variables and other variables in the model are not estimated directly, but implied by the model. There are no options to specify covariances related to latent endogenous variables in the GUI or in command notation.

Error.

Mathematically, error variables are latent exogenous variables. In `sem`, however, errors are different in that error variables have different defaults.

Errors are always named `e`. The error variable associated with observed endogenous variable `y1` has full name `e.y1`. The error variable associated with latent endogenous variable `L1` has full name `e.L1`.

In the GUI, when you create an endogenous variable, the variable's corresponding error variable immediately springs into existence. The same happens in the command language, you just do not see it. In addition, error variables automatically and inalterably have their path coefficient constrained to be 1.

Errors are treated as being uncorrelated with all other variables except as you indicate otherwise. If you want to allow the error on `y1` to be correlated with the error on `y2`, draw a curved path between `e.y1` and `e.y2` in the GUI, or include option `cov(e.y1*e.y2)` in the command language.

Finally, there is a sixth type of variable that we sometimes find convenient to talk about:

Measure, measurement.

A measure variable is an observed endogenous variable with a path from a latent variable. We introduce the word measure not as a computer term but as a convenience when communicating with humans. It is a lot easier to say that `x1` is a measure of `X` than to say that `x1` is an observed endogenous variable with a path from latent variable `X` and so, in a real sense, `x1` is a measurement of `X`.

In our measurement model,

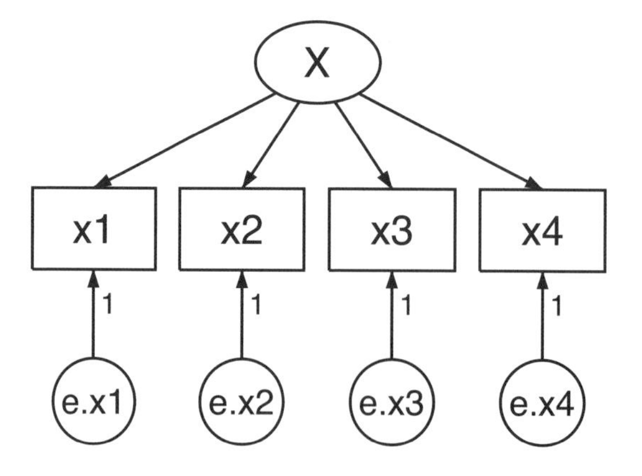

the variables are

latent exogenous:	`X`
error:	`e.x1, e.x2, e.x3, e.x4`
observed endogenous:	`x1, x2, x3, x4`

All the observed endogenous variables in this model are measures of `X`.

Constraining parameters

Constraining path coefficients to specific values

If you wish to constrain a path coefficient to a specific value, you just write the value next to the path. In our measurement error model without correlation of the residuals,

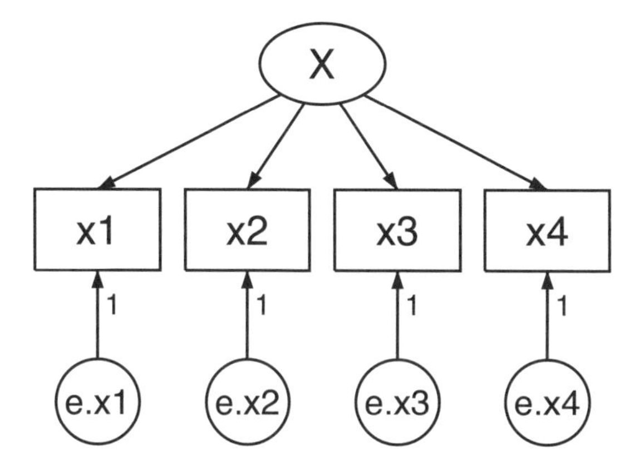

we indicate that the coefficients from `e.x1`, ..., `e.x4` are constrained to be one by placing a small 1 along the path.

We can similarly constrain any path in the model.

If we wanted to constrain $\beta_2 = 1$ in the equation

$$x_2 = \alpha_2 + X\beta_2 + e.x_2$$

we would write a 1 along the path between X and x_2. If we were instead using `sem`'s command language, we would write

```
(x1<-X) (x2<-X@1) (x3<-X) (x4<-X)
```

That is, you type an at (`@`) sign immediately after the variable whose coefficient is being constrained, and you type the value.

Constraining intercepts to specific values (suppressing the intercept)

Constraining path coefficients is common. Constraining intercepts is less so, and usually when the situation arises, you wish to constrain the intercept to zero, which is often called *suppressing the intercept.*

Although it is unusual to draw the paths corresponding to intercepts in path diagrams, they are assumed, and you could draw them if you wish. A more explicit version of our path diagram for the measurement model is

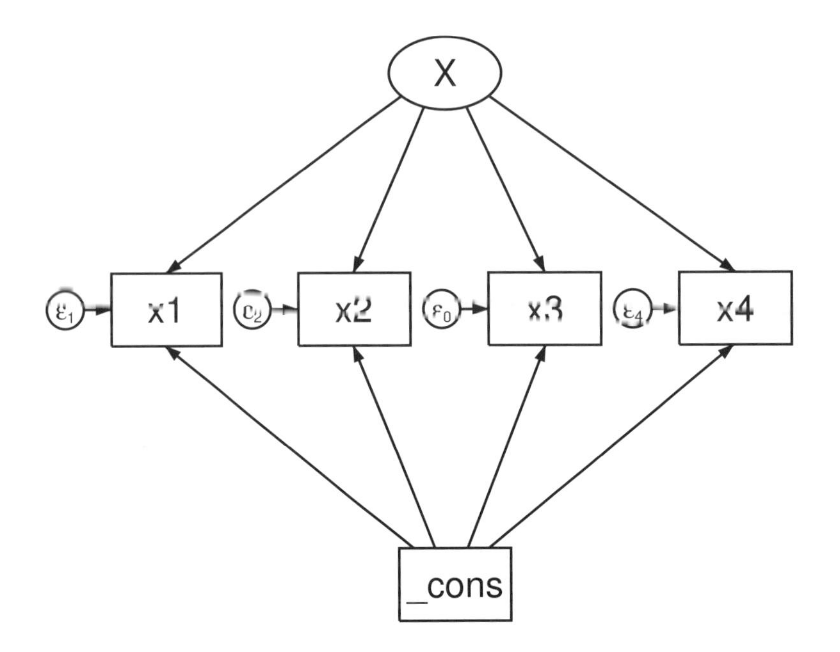

The path coefficient of `_cons` to `x1` corresponds to α_1 in

$$x_1 = \alpha_1 + X\beta_1 + e.x_1$$

and the path coefficient of `_cons` to `x2` to corresponds to α_2 in

$$x_2 = \alpha_2 + X\beta_2 + e.x_2$$

and so on.

Obviously, if you wanted to constrain a particular intercept to a particular value, you would write the value along the path. To constrain $\alpha_2 = 0$, you could draw

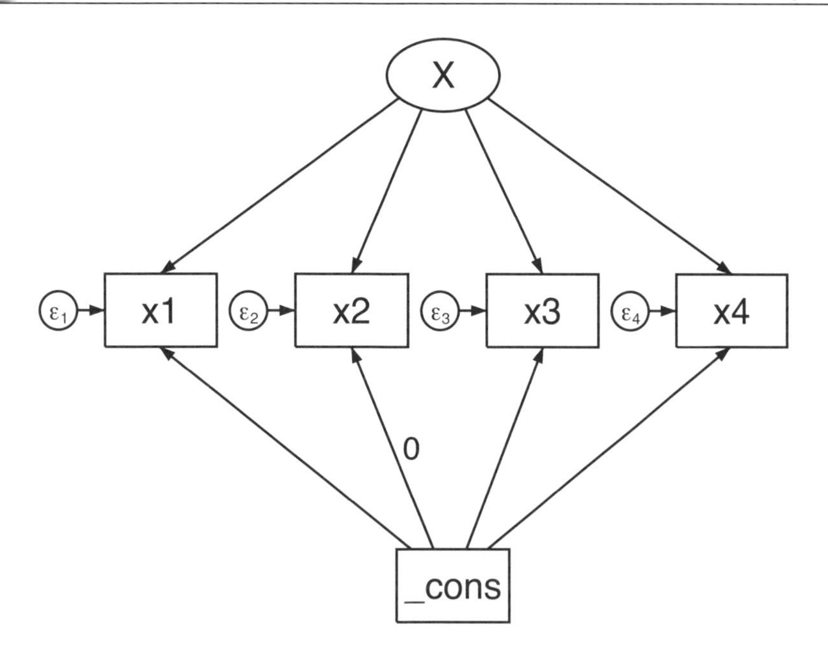

Because intercepts are assumed, you could omit drawing the paths from _cons to x1, _cons to x3, and _cons to x4:

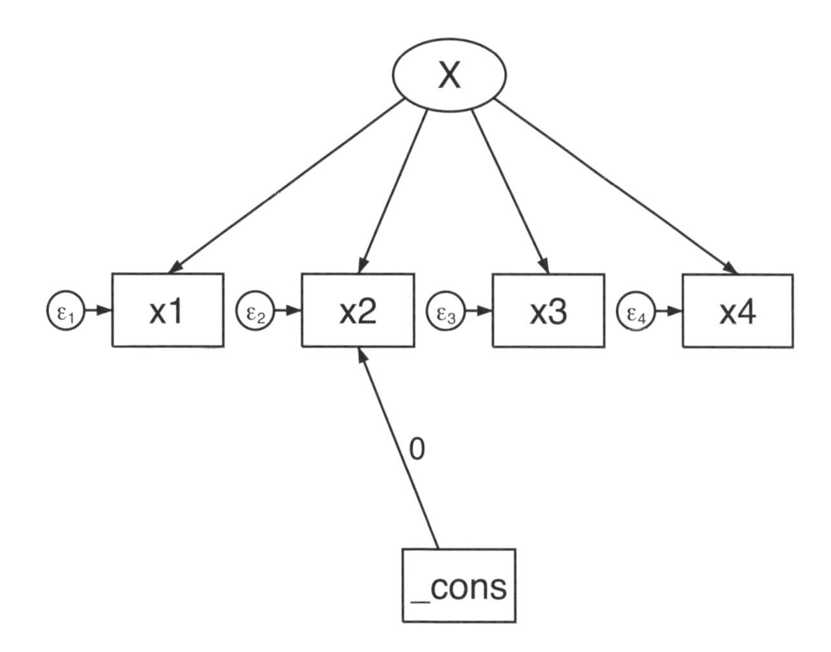

Just like the GUI, the command language assumes paths from _cons to all endogenous variables, but you could type them if you wished:

```
(x1<-X _cons) (x2<-X _cons) (x3<-X _cons) (x4<-X _cons)
```

If you wanted to constrain $\alpha_2 = 0$, you could type

```
(x1<-X _cons) (x2<-X _cons@0)  (x3<-X _cons) (x4<-X _cons)
```

or you could type

```
(x1<-X)  (x2<-X _cons@0)  (x3<-X)  (x4<-X)
```

Constraining path coefficients or intercepts to be equal

If you wish to constrain two or more path coefficients to be equal, place a symbolic name along the relevant paths:

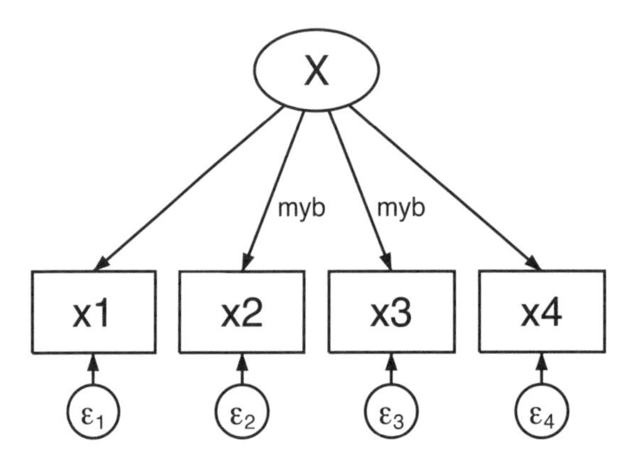

In the diagram above, we constrain $\beta_2 = \beta_3$ because we stated that $\beta_2 =$ `myb` and $\beta_3 =$ `myb`.

You follow the same approach in the command language:

```
(x1<-X) (x2<-X@myb) (x3<-X@myb) (x4<-X)
```

This works the same way with intercepts. Intercepts are just paths from `_cons`, so to constrain intercepts to be equal, you add symbolic names to their paths. In the command language, you constrain $\alpha_1 = \alpha_2$ by typing

```
(x1<-X _cons@c) (x2<-X _cons@c) (x3<-X) (x4<-X)
```

See [SEM] **example 8**.

Constraining covariances to be equal (or to specific values)

If you wish to constrain covariances, usually you will want to constrain them to be equal instead of to a specific value. If we wanted to fit our measurement model and allow correlation between `e.x2` and `e.x3` and between `e.x3` and `e.x4`, and we wanted to constrain the covariances to be equal, we could draw

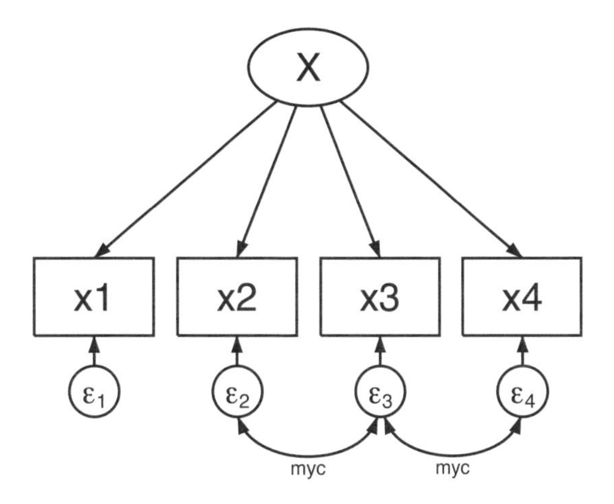

If you instead wanted to constrain the covariances to specific values, you would place the numbers along the paths in place of the symbolic names.

In the command language, covariances (curved paths) are specified using the `cov()` option. To allow covariances between `e.x2` and `e.x3` and between `e.x3` and `e.x4`, you would type

```
(x1<-X) (x2<-X) (x3<-X) (x4<-X), cov(e.x2*e.x3) cov(e.x3*e.x4)
```

To constrain the covariances to be equal, you would type

```
(x1<-X) (x2<-X) (x3<-X) (x4<-X), cov(e.x2*e.x3@myc) cov(e.x3*e.x4@myc)
```

Constraining variances to specific values (or to be equal)

Variances are like covariances except that in path diagrams drawn by some authors, variances curve back on themselves. In `sem`'s GUI, variances appear inside or beside the box or circle. Regardless of how they appear, variances may be constrained to normalize latent variables, although normalization is handled by `sem` automatically (something we will explain in *How sem solves the problem for you* under *Identification 2: Normalization constraints (anchoring)* below).

In the GUI, you constrain variances by clicking on the variable and using the lock box to specify the value, which can be a number or a symbol. In the command language, variances are specified using the `var()` option as we will explain below.

Let's assume that you wanted to normalize the latent variable `X` by constraining its variances to be 1. You could do that by drawing

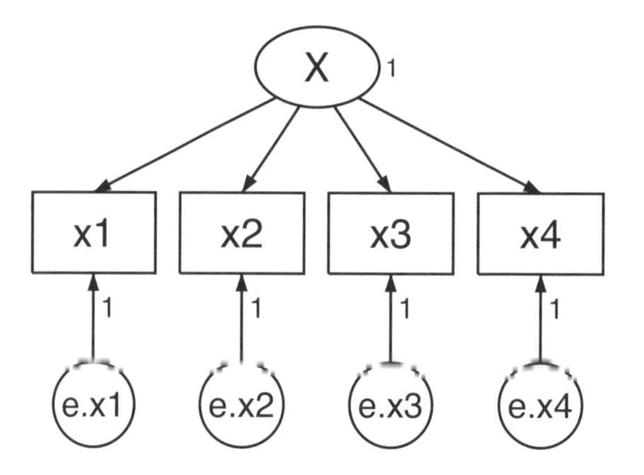

In the command language, we would specify this model as

```
(x1<-X) (x2<-X) (x3<-X) (x4<-X), var(X@1)
```

Constraining latent exogenous variables to have unit variance as an identifying restriction may be desirable when you wish simultaneously to constrain their correlations with other latent exogenous variables. `sem` allows you to constrain covariances, not correlations. Covariances are equal to correlations when variances are one.

It may happen in more complicated models that you wish to constrain variances of latent exogenous variables to be equal. You can do that by specifying a symbolic name.

Identification 1: Substantive issues

Not all models are identified

Just because you can draw the path diagram for a model, or equivalently, write its equations, or write it in Stata's command syntax, does not mean the model is identified. Identification refers to the conceptual constraints on parameters of a model that are required for the model's remaining parameters to have a unique solution. A model is said to be unidentified if these constraints are not supplied. These constraints are of two types: substantive constraints and normalization constraints. We will begin by discussing substantive constraints because that is your responsibility; the software provides normalization constraints automatically.

How to count parameters

If your model has K observed variables, then your data contain $K(K+1)/2$ second-order moments, and thus p, the number of parameters based on second-order moments that can be estimated, cannot exceed $K(K+1)/2$.

Every path in your model contributes 1 to p unless the parameter is constrained to a specific value, and then it does not count at all. If two parameters are constrained to be equal, the two parameters count as one. In counting p, you must remember to count the curved paths from latent variables back to themselves, which is to say, the variances. Just counting the number of parameters can be challenging. And even if $p \leq K(K+1)/2$, your model may not be identified. Identification depends not only on the number of paths but also on their location.

Books have been written on this subject, and we will refer you to them. A few are Bollen (1989), Brown (2006), Kline (2011), and Kenny (1979). We will refer you to them, but do not be surprised if they refer you back to us. Brown (2006, 202) writes, "Because latent variable software programs are capable of evaluating whether a given model is identified, it is often most practical to simply try to estimate the solution and let the computer determine the model's identification status." That is not bad advice.

What happens when models are unidentified

So what happens when you attempt to fit an unidentified model? In some cases, `sem` will tell you that your model is unidentified, because `sem` applies the counting rule. If your model is unidentified for more subtle substantive reasons, however, `sem` will iterate forever, reporting the same criterion value (such as log likelihood) and saying "not concave" over and over again. The output looks like this:

```
Iteration 50:   log likelihood = -337504.44  (not concave)
Iteration 51:   log likelihood = -337504.44  (not concave)
Iteration 52:   log likelihood = -337504.44  (not concave)
.
.
.
Iteration 101:   log likelihood = -337504.44  (not concave)
.
.
.
```

Observing periods of the "not concave" message is not concerning, so do not overreact at the first occurrence. Become concerned when you see "not concave" and the criterion value is not changing, and even then, stay calm for a short time because the value might be changing in digits you are not seeing. If the iteration log continues to report the same value several times, however, press *Break*. Your model is probably not identified.

How to diagnose and fix the problem

You must find and fix the problem. A useful trick is to rerun the model, but this time, specify `sem`'s `iterate(#)` option, where # is large enough to reach the "not concave" messages with constant criterion value. `sem` will iterate that many times and then report the results it has at that point. Look at the output for missing standard errors. Those parameters are unidentified, and you need to think about changing your model so that they become identified or placing constraints on them.

Identification 2: Normalization constraints (anchoring)

Models with latent variables require normalization constraints because latent variables have no natural scale. If constraints are not provided, the model will appear to the software just like a model with a substantive lack of identification; the estimation routine will iterate forever and never arrive at a solution. The `sem` command automatically provides normalization constraints.

Below we explain why the normalization constraints are required, which normalization constraints `sem` automatically supplies, how to override those automatic choices, and how to substitute your own constraints should you desire.

Why the problem arises

Imagine a latent variable for propensity to be violent. Your imagination might supply a scale that ranges from 0 to 1 or 1 to 100 or over other values, but regardless, the scale you imagine is arbitrary in that one scale works as well as another.

Scales have two components: mean and variance. If you imagine a latent variable with mean 0 and your colleague imagines the same variable with mean 100, the difference can be accommodated in the parameter estimates by an intercept changing by 100. If you imagine a standard deviation of 1 (variance $1^2 = 1$) and your colleague imagines a standard deviation of 10 (variance $10^2 = 100$), the difference can be accommodated by a path coefficient differing by a multiplicative factor of 10. You might measure an effect as being 1.1 and then your colleague would measure the same effect as being 0.11, but either way you both will come to the same substantive conclusions.

How the problem would manifest itself

The problem is that different scales all work equally well, and the software will iterate forever, jumping from one scale to another.

Another way of saying that means and variances of latent variables are arbitrary is to say that they are unidentified. That's important because if you do not specify the scale you have in mind, results of estimation will look just like substantive lack of identification.

`sem` will iterate forever and never arrive at a solution.

How sem solves the problem for you

You usually do not need to worry about this problem because `sem` solves it for you. `sem` solves the unidentified scale problem by

1. Assuming that all latent exogenous variables have mean 0.
2. Assuming that all latent endogenous variables have intercept 0.
3. Setting the coefficients on paths from latent variables to the first observed endogenous variable to be 1.
4. Setting the coefficients on paths from latent variables to the first latent endogenous variable to be 1 if rule 3 does not apply—if the latent variable is measured by other latent variables only.

Rules 3 and 4 are also known as the unit-loading rules. The variable to which the path coefficient is set to 1 is said to be the anchor for the latent variable.

Applying those rules to our measurement model, when we type

```
(X->x1) (X->x2) (X->x3) (X->x4)
```

`sem` acts as if we typed

```
(X@1->x1) (X->x2) (X->x3) (X->x4), means(X@0)
```

The above four rules are sufficient to provide a scale for latent variables for all models.

Overriding sem's solution

sem automatically applies rules 1 through 4 to produce normalization constraints. There are, however, other normalization constraints that would work as well. In what follows, we will assume that you are well versed in deriving normalization constraints and just want to know how to bend sem to your will.

Before you do this, however, let us warn you that substituting your normalization rules for sem's defaults can result in more iterations being required to fit your model. Yes, one set of normalization constraints are as good as the next, but sem's starting values are based on its default normalization rules, and that means that when you substitute your rules for sem's, the required number of iterations sometimes increases.

Let's return to the measurement model:

```
(X->x1) (X->x2) (X->x3) (X->x4)
```

As we said previously, type the above and sem acts as if you typed

```
(X@1->x1) (X->x2) (X->x3) (X->x4), means(X@0)
```

If you wanted to assume instead that the mean of X was 100, you could type

```
(X->x1) (X->x2) (X->x3) (X->x4), means(X@100)
```

The means() option allows you to specify mean constraints, and you may do so for latent or observed variables.

Let's leave the mean at 0 and specify that we instead want to constrain the second path coefficient to be 1. Type

```
(X->x1) (X@1->x2) (X->x3) (X->x4)
```

We did not have to tell sem not to constrain X->x1 to have coefficient 1. We just specified that we wanted to constrain X->x2 to have coefficient 1. sem takes all the constraints that you specify and then it adds normalization constraints to identify the model. If what you have specified is sufficient, sem does not add its default normalization constraints because they are no longer necessary.

Obviously, if we wanted to constrain the mean to 100 and the second rather than the first path coefficient to 1, we could type

```
(X->x1) (X@1->x2) (X->x3) (X->x4), means(X@100)
```

If we wanted to constrain the standard deviation of X to be 10 instead of constraining a path coefficient to be 1, we could type

```
(X->x1) (X->x2) (X->x3) (X->x4), means(X@100) var(X@100)
```

Standard deviations are specified in variance units when you use the var() option.

Starting values

It can be devilishly difficult for software to obtain results for SEM models because SEM models are based on second-order moments. Sometimes the software fails. You need to learn to identify when the software is in trouble so that you can press the *Break* key and end the failed attempt to find a solution. You then need to learn how to specify better starting values—and to identify the parameters that need them—so that you can restart the estimation process.

We provide step-by-step instructions for dealing with convergence problems in [SEM] **sem**. Go there the day you face the problem. Right now, however, let's examine the issues.

What happens when starting values are not good enough

SEM models are fit by an iterative process. That process is kicked off by the software choosing a set of values for the parameters that is good enough for the process to iterate to a better value and so, by repeatedly iterating, find the solution. Sometimes those starting values are not good enough, and you will see

```
. sem ...
Variables in structural equation model
  (output omitted )
Fitting target model:
initial values not feasible
r(1400);
```

or you will see

```
. sem ...
Variables in structural equation model
  (output omitted )
Fitting target model:
     Iteration  1:  log likelihood = ...
     .
     .
     .
     Iteration 50:  log likelihood = -337504.44   (not concave)
     Iteration 51:  log likelihood = -337503.52   (not concave)
     Iteration 52:  log likelihood = -337502.13   (not concave)
     .
     .
     .
     Iteration 101: log likelihood = -337400.69   (not concave)
--Break--
r(1);
```

In the first case, we received an error message, "initial values not feasible". In the second case, we saw we were in trouble and pressed *Break*. We saw we were in trouble because the log-likelihood value was increasing so slowly.

What to do when starting values are not good enough

Whether you see the "initial values not feasible" error or you pressed *Break*, you need to look at the parameter estimates that `sem` currently has.

If you saw the "initial values not feasible" error, retype your `sem` command and add the `trace` option:

```
. sem ..., trace
```

If you instead pressed *Break* to stop the iterations, add the `iterate(3)` option:

```
. sem ..., iterate(3)
```

Either way, you will now have the parameter estimates in front of you. Scan them and find the one or more that are impossible (for example, negative variance) or ridiculous (for example, very small but still positive variance). The problem will usually be with the variance of a latent exogenous variable, or of an error associated with a latent endogenous variable, going to zero.

Say that you suspect the problem is with the variance of variable `F` going to zero. You need to give that variable a larger starting value, which you can do by typing

```
. sem ..., var(e.F, init(1))
```

or

```
. sem ..., var(e.F, init(2))
```

or

```
. sem ..., var(e.F, init(10))
```

We recommend choosing a value for the variance that is larger than you believe is necessary. To obtain an estimate of the variance,

1. If the variable is observed, use `summarize` to obtain the summary statistics for the variable, square the reported standard deviation, and then increase that by, say, 20%.
2. If the variable is latent, use `summarize` to obtain a summary of the latent variable's anchor variable and then follow the same rule: use `summarize`, square the reported standard deviation, and then increase that by 20%. (The anchor variable is the variable whose path is constrained to have coefficient 1.)
3. If the variable is latent and has paths only to other latent variables so that its anchor variable is itself latent, follow the anchor paths to an observed variable and follow the same rule: use `summarize`, square the reported standard deviation, and then increase that by 20%.

Do not dismiss the possibility that the bad starting values concern estimated parameters other than variances of latent exogenous or error variables, although variances of those kinds of variables is the usual case. Covariances are rarely the problem because covariances can take on any value and, whether too small or too large, usually get themselves back on track as the iterative process proceeds. If you need to specify an initial value for a covariance, the syntax is

```
. sem ..., cov(e.F*e.G, init(-25))
```

Substitute for -25 the value you consider reasonable.

The other possibility is that a path coefficient needs a better starting value, which is as unlikely as a covariance being the problem, and for the same reasons. To set the initial value of a path coefficient, add the `init()` option where the path is specified. Say the original `sem` command included `y<-x1`:

```
. sem ... (y<-x1 x2) ...
```

If you wanted to set the initial value of the path from `x1` to 3, modify the command to read

```
. sem ... (y<-(x1, init(3)) x2) ...
```

Distinguishing poor starting values from lack of identification

The problem of poor starting values can produce output that looks very much like the problem of identification unless you look closely. If the problem is poor starting values, the criterion function being optimized (for example, the log likelihood) will improve iteration by iteration; it will just improve slowly. If the problem is lack of identification, the criterion value will remain constant. In addition, when you look at results, if the problem is starting values, it is likely that a variance estimate is heading toward zero, or is zero, or negative. If the problem is with an absurd path coefficient, that is more likely to be caused by lack of identification.

Also see

[SEM] **intro 2** — Learning the language: Path diagrams and command language

[SEM] **intro 4** — Tour of models

[SEM] **sem path notation** — Command syntax for path diagrams

[SEM] **sem option covstructure()** — Specifying covariance restrictions

Title

intro 4 — Tour of models

Description

Below is a sampling of structural equation models that can be fit by `sem`.

Remarks

If you have not read [SEM] **intro 2**, please do so. You need to speak the language. We also recommend reading [SEM] **intro 3**, but that is not required.

Now that you speak the language, we can start all over again and take a look at some of the classic models that `sem` can fit.

Remarks are presented under the following headings:

Single-factor measurement models

See [SEM] **example 1**.

A single-factor measurement model is

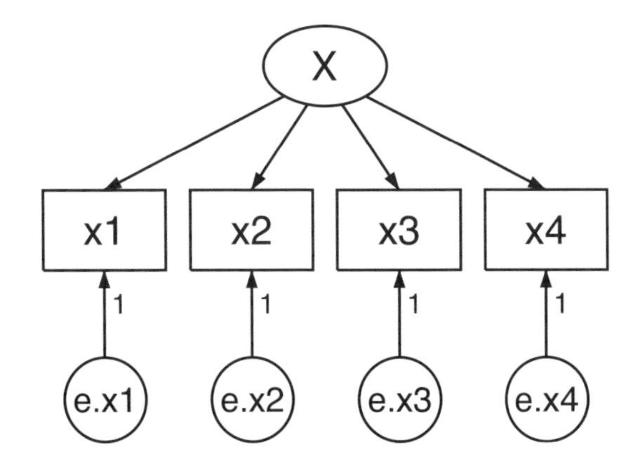

The model can be written in Stata command language as

```
(x1<-X) (x2<-X) (x3<-X) (x4<-X)
```

or as

```
(x1 x2 x3 x4<-X)
```

or as

```
(X->x1 x2 x3 x4)
```

or in other ways. All the equivalent ways really are equivalent; no subtle differences will subsequently arise according to your choice.

The measurement model plays an important role in many other SEMs dealing with the observed inputs and the observed outputs:

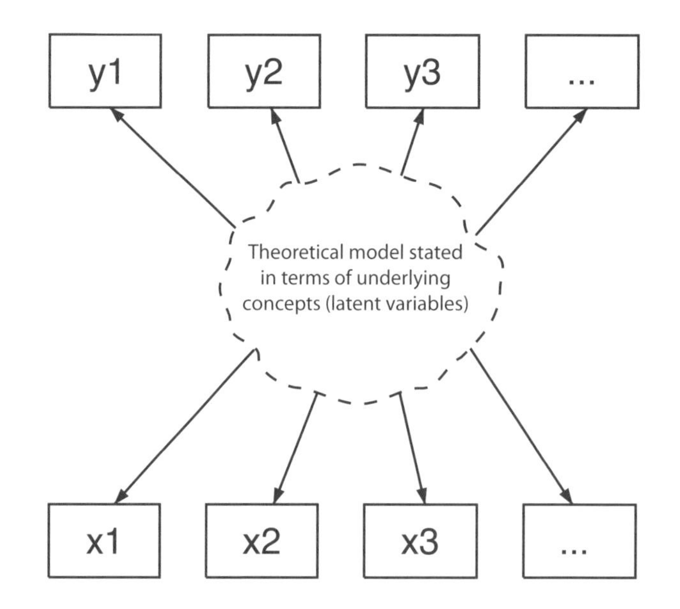

Because the measurement model is so often joined with other models, it is common to refer to the coefficients on the paths from latent variables to observable endogenous variables as the measurement coefficients and their intercepts as the measurement intercepts. The intercepts are usually not shown in path diagrams. The other coefficients and intercepts are those not related to the measurement issue.

The measurement coefficients are often referred to as loadings.

Multiple-factor measurement models

See [SEM] **example 3**.

A two-factor measurement model is two one-factor measurement models with possible correlation between the factors:

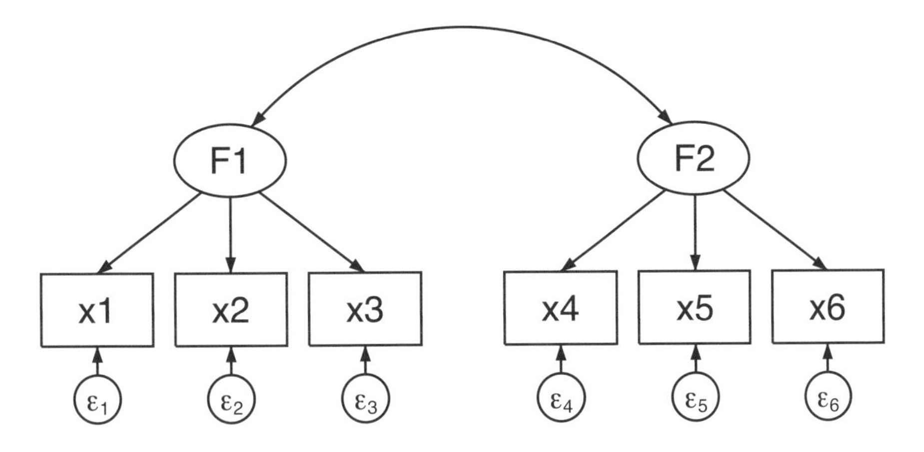

To obtain a correlation between F1 and F2, we drew a curved path.

The model can be written in Stata command language as

```
(F1->x1) (F1->x2) (F1->x3) (F2->x4) (F2->x5) (F2->x6)
```

In the command language, we do not have to include the cov(F1*F2) option because, by default, sem assumes that exogenous latent variables are correlated with each other.

This model can also be written in any of the following ways:

```
(F1->x1 x2 x3) (F2->x4 x5 x6)
```

or

```
(x1 x2 x3<-F1) (x4 x5 x6<-F2)
```

or

```
(x1<-F1) (x2<-F1) (x3<-F1) (x4<-F2) (x5<-F2) (x6<-F2)
```

CFA models

See [SEM] **example 5**.

The measurement models just shown are also known as confirmatory factor analysis (CFA) models because they can be analyzed using CFA.

In the single-factor model, after estimation, you might want to test that all the indicators have significant loadings by using test; see [SEM] **test**. You might also want to test whether the correlations between the errors should have been included in the model by using estat mindices; see [SEM] **estat mindices**.

In the multiple-factor measurement model, you might want to test that any of the omitted paths should in fact be included in the model. The omitted paths in the two-factor measurement model above were F1 → x4, F1 → x5, F1 → x6, and F2 → x1, F2 → x2, F2 → x3. estat mindices will perform these tests.

We show other types of CFA models below.

Structural models 1: Linear regression

See [SEM] **example 6**.

Different authors define the meaning of structural models in different ways. Bollen (1989, 4) defines a structural model as the parameters being not just of a descriptive nature of association but instead of a casual nature. By that definition, the measurement models above could be structural models, and so could the linear regression below.

Others define structural models as models having paths reflecting causal dependencies between endogenous variables and thus would exclude the measurement model and linear regression. We will show you a "true" structural model in the next example.

An example of a linear regression would be

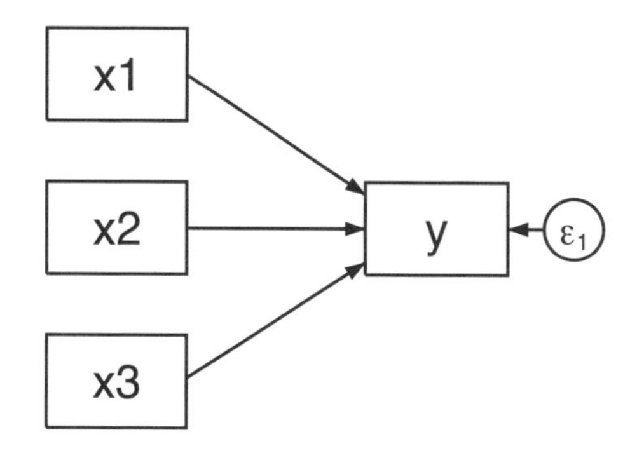

The model above can be written in Stata command language as

```
(y <- x1 x2 x3)
```

When you estimate a linear regression using `sem`, you obtain the same point estimates as you would with `regress` and the same standard errors up to a degree-of-freedom adjustment applied by `regress`.

Structural models 2: Dependencies between endogenous variables

See [SEM] **example 7**.

An example of a structural model having paths between endogenous variables would be

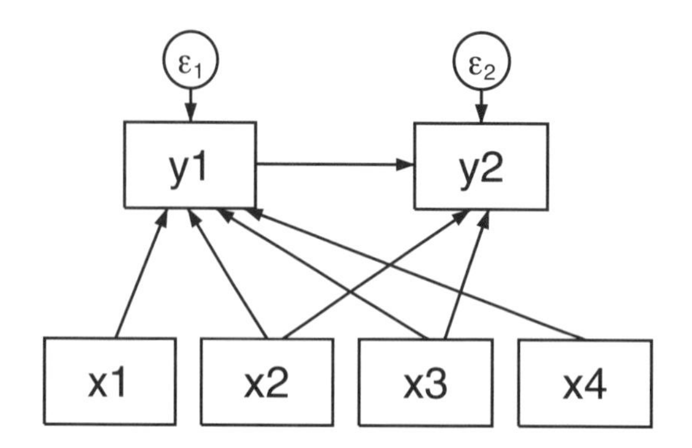

The model above can be written in Stata command language as

```
(y1 <- x1 x2 x3 x4) (y2 <- y1 x2 x3)
```

In this example, all inputs and outputs are observed and the errors are assumed to be uncorrelated. In these kinds of models, it is common to allow correlation between errors:

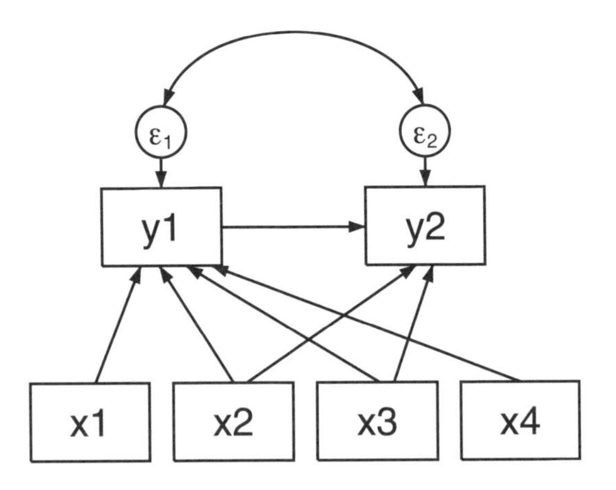

The model above can be written in Stata command language as

```
(y1 <- x1 x2 x3 x4) (y2 <- y1 x2 x3), cov(e.y1*e.y2)
```

This structural model is said to be overidentified. If we omitted y1 ← x4, the model would be just identified. If we also omitted y1 ← x1, the model would be unidentified.

When you fit the above model using sem, you obtain slightly different results from those you would obtain with ivregress liml. This is because sem with default method(ml) produces full information maximum likelihood rather than limited-information maximum likelihood results.

Structural models 3: Unobserved inputs, outputs, or both

See [SEM] **example 9**.

Perhaps in a structural model such as

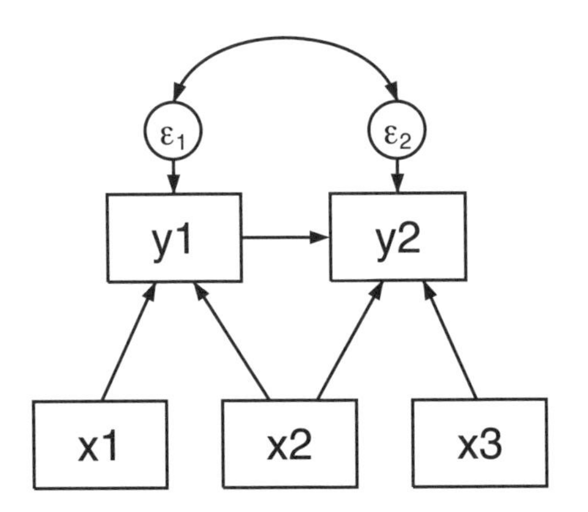

the inputs x1, x2, and x3 are concepts and thus are not observed. Assume that we have measurements for them. We can join this structural model example with a three-factor measurement model:

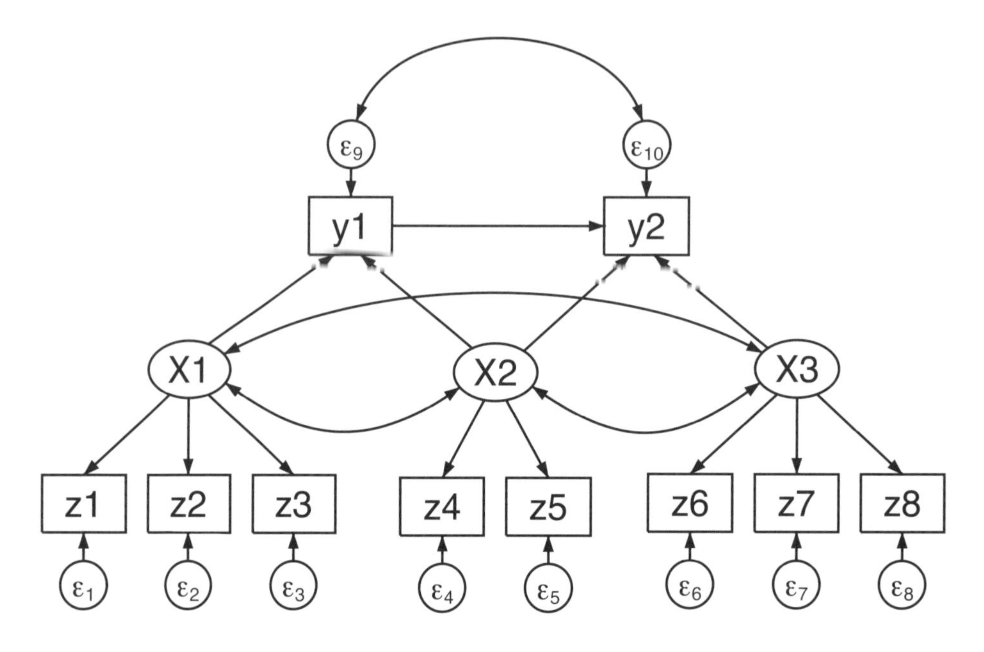

Note the curved arrows denoting correlation between the pairs of X1, X2, and X3. In the previous path diagram, we had no such arrows between the variables, yet we were still assuming that they were there. In sem's path diagrams, correlation between exogenous observed variables is assumed and need not be explicitly shown. When we changed observed variables x1, x2, and x3 to be the latent variables X1, X2, and X3, we needed to show explicitly the correlations we were allowing. Correlation between latent variables is not assumed unless shown.

This model can be written in Stata command syntax as follows:

```
(y1<-X1 X2) (y2<-y1 X2 X3)      ///
(X1->z1 z2 z3)                  ///
(X2->z4 z5)                     ///
(X3->z6 z7 z8),                 ///
             cov(e.y1*e.y2)
```

We did not include the `cov(X1*X2 X1*X3 X2*X3)` option, although we could have. In the command language, exogenous latent variables are assumed to be correlated with each other. If we did not want X2 and X3 to be correlated, we would need to include the `cov(X2*X3@0)` option.

We changed x1, x2, and x3 to be X1, X2, and X3. In command syntax, variables beginning with a capital letter are assumed to be latent. Alternatively, we could have left the names in lowercase and specified the identities of the latent variables:

```
(y1<-x1 x2) (y2<-y1 x2 x3)      ///
(x1->z1 z2 z3)                  ///
(x2->z4 z5)                     ///
(x3->z6 z7 z8),                 ///
             cov(e.y1*e.y2)     ///
             latent(x1 x2 x3)
```

Just as we have joined an observed structural model to a measurement model to handle unobserved inputs, we could join the above model to a measurement model to handle unobserved y1 and y2.

Structural models 4: MIMIC

See [SEM] **example 10**.

MIMIC stands for multiple indicators and multiple causes. An example of a MIMIC model is

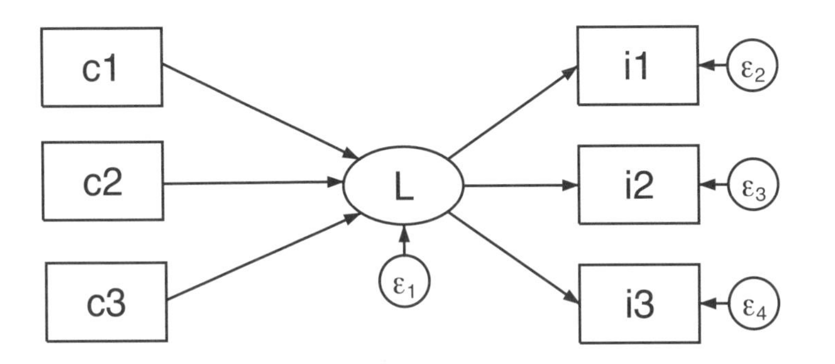

In this model, the observed causes c1, c2, and c3 determine latent variable L, and L in turn determines observed indicators i1, i2, and i3.

This model can be written in Stata command syntax as

```
(i1 i2 i3 <- L) (L <- c1 c2 c3)
```

Structural models 5: Seemingly unrelated regression (SUR)

See [SEM] **example 12**.

Seemingly unrelated regression is like having two or more separate linear regressions but allowing the errors to be correlated.

An example of a seemingly unrelated regression (SUR) model is

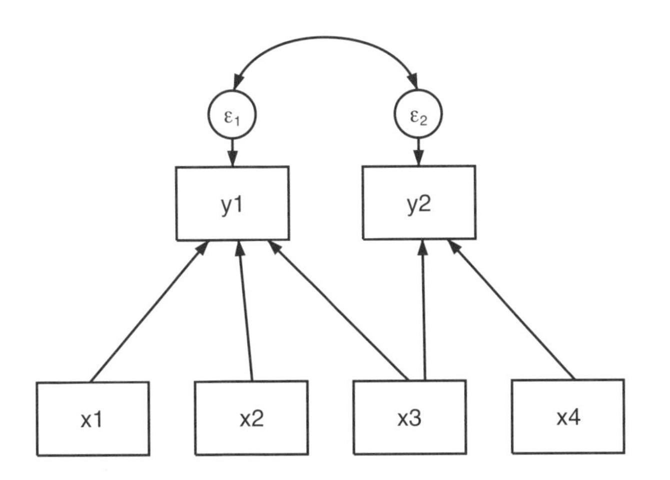

The model above can be written in Stata command syntax as

```
(y1 <- x1 x2 x3) (y2 <- x3 x4), cov(e.y1*e.y2)
```

In this example, the two regressions shared a common exogenous variable, x3. That is not necessary. Or, they could share more variables. If they shared all variables, results would be the same as estimating multivariate regression, shown in the next example.

When you estimate an SUR using sem, you obtain the same point estimates as you would with sureg if you specify sureg's isure option, which causes sureg to iterate until it obtains the maximum likelihood result. Standard errors will be different. If the model has exogenous variables only on the right-hand side, standard errors will be asymptotically identical and, although the standard errors are different in finite samples, there is no reason to prefer one set over the other. If the model being fit is recursive, standard errors produced by sem are better than those from sureg, both asymptotically and in finite samples.

Structural models 6: Multivariate regression

See [SEM] **example 12**, even though the example is of SUR. Multivariate regression is a special case of SUR.

A multivariate regression is just an SUR where the different dependent variables share the same exogenous variables:

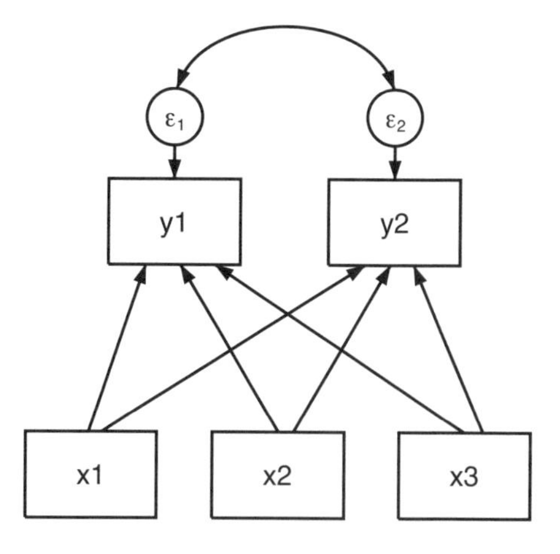

The model above can be written in Stata command syntax as

```
(y1 y2 <- x1 x2 x3), cov(e.y1*e.y2)
```

When you estimate a multivariate regression using sem, you obtain the same point estimates as you would with mvreg and the same standard errors up to a multiplicative $\sqrt{(N-p-1)/N}$ degree-of-freedom adjustment applied by mvreg.

Correlations

See [SEM] **example 16**.

We are all familiar with correlation matrices of observed variables, such as

	x1	x2	x3
x1	1.0000		
x2	0.7700	1.0000	
x3	−0.0177	−0.2229	1.0000

or covariances matrices, such as

	x1	x2	x3
x1	662.172		
x2	62.5157	9.95558	
x3	−0.769312	−1.19118	2.86775

These results can be obtained from `sem`. The path diagram for the model is

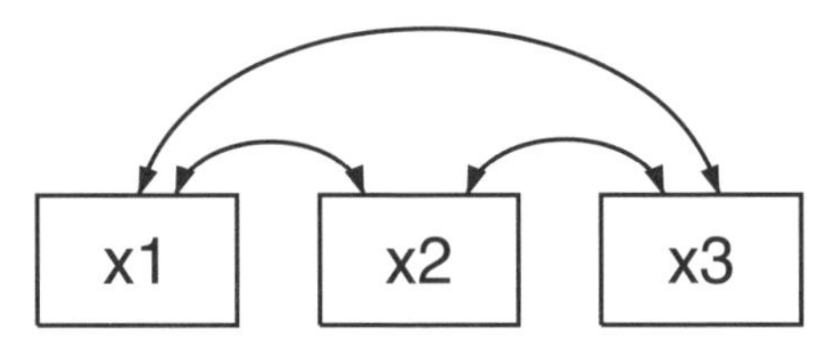

We could just as well leave off the curved paths because `sem` assumes them among observed exogenous variables:

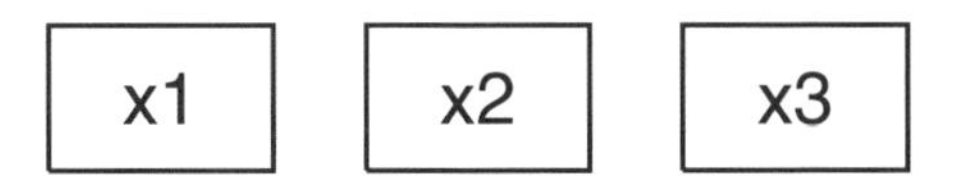

Either way, this model can be written in Stata command syntax as

```
(<- x1 x2 x3)
```

That is, we simply omit specifying the target of the path, the endogenous variable.

If we fit the model, we will obtain the covariance matrix by default. `correlate` with the `covariance` option produces covariances that are divided by $N-1$ rather than N. To match this covariance exactly, you need to specify the `nm1` option, which we can do in the command language by typing

```
(<- x1 x2 x3), nm1
```

If we want correlations rather than covariances, we ask for them by specifying the `standardized` option:

```
(<- x1 x2 x3), nm1 standardized
```

An advantage of obtaining correlation matrices from `sem` rather than from `correlate` is that you can perform statistical tests on the results, such as that the correlation of `x1` and `x3` is equal to the correlation of `x2` and `x3`.

If you are willing to assume joint normality of the variables, you can obtain more efficient estimates of the correlations in the presence of missing-at-random data by specifying the `method(mlmv)` option.

Higher-order CFA models

See [SEM] **example 15**.

Sometimes observed values measure traits or other aspects of latent variables, so we insert a new layer of latent variables to reflect those traits or aspects. We have measurements—say, `x1`, ..., `x6`—all reflecting underlying factor `F`, but `x1` and `x2` measure one trait of `F`, `x3` and `x4` measure another trait, and `x5` and `x6` measure yet another trait. This model can be drawn as

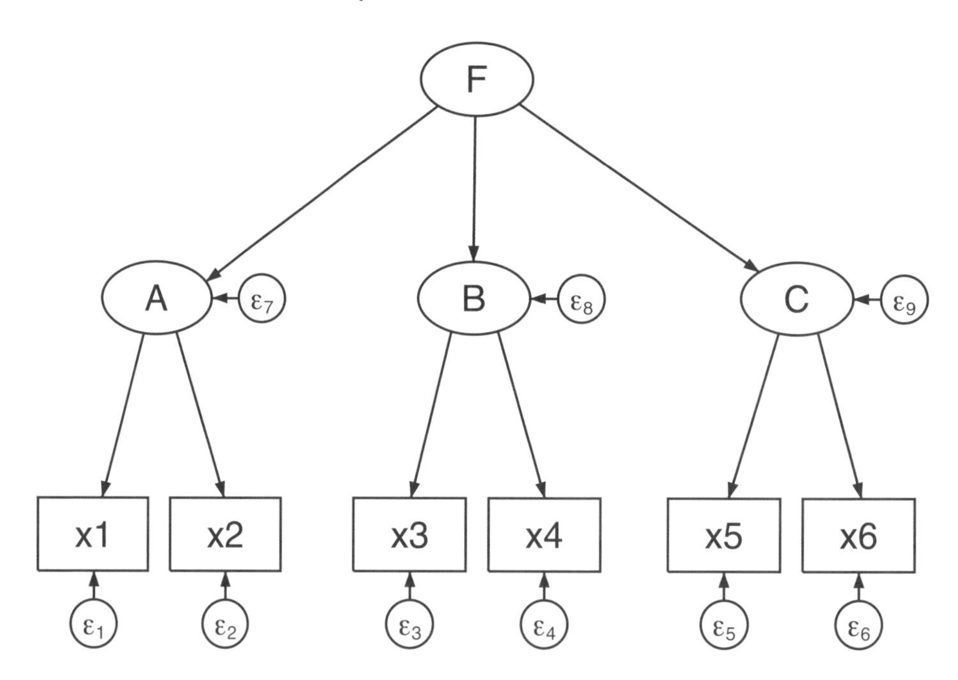

The above model can be written in command syntax as

```
(A->x1 x2) (B->x3 x4) (C->x5 x6) (A B C <- F)
```

Correlated uniqueness model

See [SEM] **example 17**.

Sometimes observed values are correlated just because of how the data are collected. Imagine we have factor `T1` representing a trait and measurements `x1` and `x4`. Perhaps `T1` is aggression, and `x1` is self reported, and `x4` is reported by the spouse. Imagine we also have another factor, `T2`, and measurements `x2` and `x5`; `x2` is self reported, and `x5` is reported by the spouse. It would not be unlikely that `x1` and `x2` are correlated and that `x4` and `x5` are correlated. That is exactly what the correlated uniqueness model assumes:

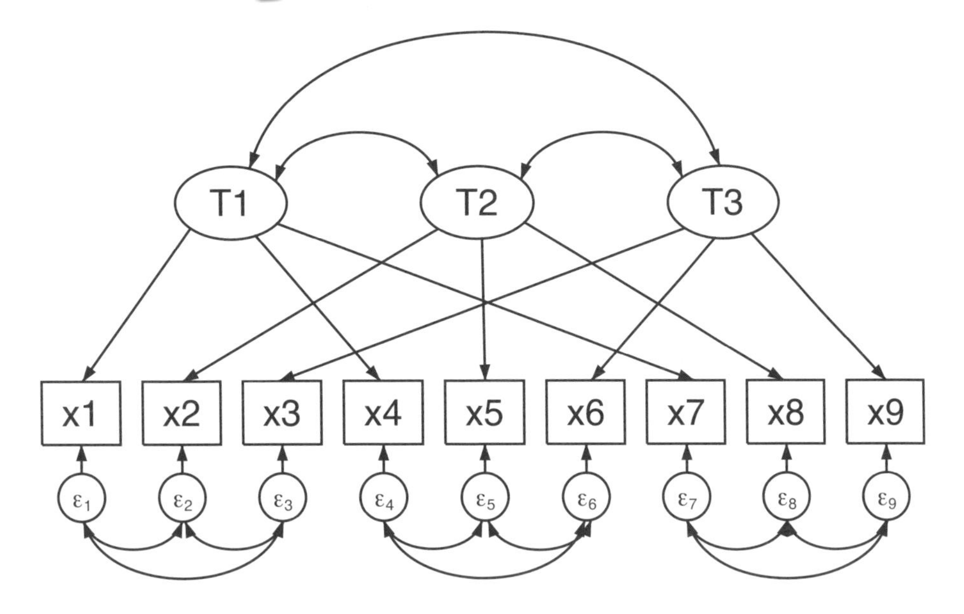

Data that exhibit this kind of pattern are known as multitrait–multimethod (MTMM) data. Researchers historically looked at the correlations, but SEM allows us to fit a model that incorporates the correlations.

The above model can be written in Stata command syntax as

```
(T1 -> x1 x4 x7)                                          ///
(T2 -> x2 x5 x8)                                          ///
(T3 -> x3 x6 x9),                                         ///
          cov(e.x1*e.x2  e.x1*e.x3  e.x2*e.x3)            ///
          cov(e.x4*e.x5  e.x4*e.x6  e.x5*e.x6)            ///
          cov(e.x7*e*x8  e.x7*e.x9  e.x8*e.x9)
```

An alternative way to type the above is to use the `covstructure()` option, which we can abbreviate as `covstruct()`:

```
(T1 -> x1 x4 x7)                                          ///
(T2 -> x2 x5 x8)                                          ///
(T3 -> x3 x6 x9),                                         ///
          covstruct(e.x1 e.x2 e.x3, unstructured)         ///
          covstruct(e.x4 e.x5 e.x6, unstructured)         ///
          covstruct(e.x7 e.x8 e.x9, unstructured)
```

Unstructured means that the listed variables have covariances. Specifying blocks of errors as unstructured would save typing if there were more variables in each block.

Latent growth models

See [SEM] **example 18**.

A latent growth model is a variation on the measurement model. In our measurement model examples, we have assumed four observed measurements of underlying factor `X`: `x1`, `x2`, `x3`, and `x4`. In the command language, which saves paper, we can write this as

```
(X->x1) (X->x2) (X->x3) (X->x4)
```

Let's assume that the observed values are collected over time. `x1` is observed at time 0, `x2` at time 1, and so on. It thus may be more reasonable to assume that the observed values represent a base value and a growth modeled with a linear trend. Thus we might write the model as

```
(B@1 L@0 -> x1)                    ///
(B@1 L@1 -> x2)                    ///
(B@1 L@2 -> x3)                    ///
(B@1 L@3 -> x4),                   ///
                 noconstant
```

which is to say, the equations are

$$x_1 = B + 0L + e.x_1$$
$$x_2 = B + 1L + e.x_1$$
$$x_3 = B + 2L + e.x_1$$
$$x_4 = B + 3L + e.x_1$$

The path diagram for the model is

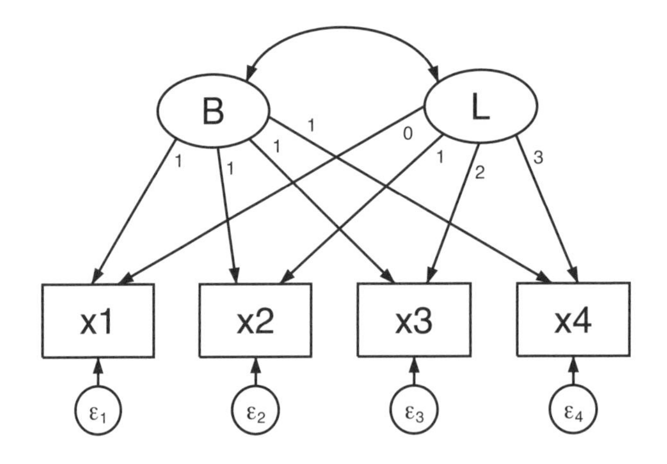

In evaluating this model, it is useful to review the means of the latent exogenous variables. In most models, latent exogenous variables have mean 0 and the means are thus uninteresting. `sem` usually constrains latent exogenous variables to have mean 0 and does not report that fact.

In this case, however, we ourselves have placed constraints, and thus the means are identified and in fact are an important point of the exercise. We must tell `sem` not to constrain the means of the two latent exogenous variables `B` and `L`, which we do with the `means()` option:

```
(B@1 L@0 -> x1)                    ///
(B@1 L@1 -> x2)                    ///
(B@1 L@2 -> x3)                    ///
(B@1 L@3 -> x4),                   ///
                 noconstant means(B L)
```

We must similarly specify the `means()` option when using the GUI.

Models with reliability

See [SEM] **example 24**.

A typical solution for dealing with variables measured with error is to find multiple measurements and to use those measurements to develop a latent variable. See, for example, *Single-factor measurement models* and *Multiple-factor measurement models* above.

When the reliability of the variables is known—reliability is measured as the fraction of variances that is not due to measurement error—another approach is available. This approach can be used in place of or in addition to the use of multiple measurements.

See [SEM] **sem option reliability()**.

Also see

[SEM] **intro 3** — Substantive concepts

[SEM] **intro 6** — Postestimation tests and predictions

[SEM] **example 1** — Single-factor measurement model

Title

intro 5 — Comparing groups

Description

You can compare groups—compare males with females, compare age group 1 with age group 2 with age group 3, and so on—with respect to any SEM model. Said more technically, any model fit by `sem` can be simultaneously estimated for different groups with some parameters constrained to be equal across groups and others allowed to vary, and those estimates can be used to perform statistical tests for comparing the groups.

Remarks

Remarks are presented under the following headings:

The generic SEM model
Fitting the model for different groups of the data
Which parameters vary by default, and which do not
Specifying which parameters are allowed to vary in broad, sweeping terms
Adding constraints for path coefficients across groups
Adding constraints for means, variances, or covariances across groups
Adding constraints for some groups but not others
Adding paths for some groups but not others
Relaxing constraints

The generic SEM model

In [SEM] **intro 4**, we noted that measurement models are often joined with other SEM models to produce

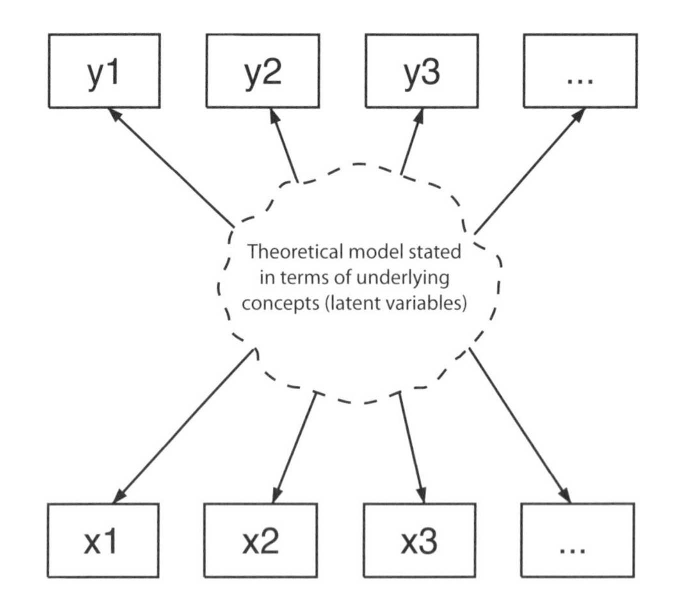

This can be written in the command syntax as

```
(Y1->...) (Y2->...)           ///
(...)                         /// theoretical model stated in terms of
(...)                         /// underlying concepts (latent variables)
(...)                         ///
(X1->...) (X2->...)
```

where the middle part is the theoretical model stated in terms of underlying concepts Y1, Y2, X1, and X2. However we write the model, we are assuming that

1. the unobserved X1 and X2 are measured by observed variables x1, x2, ...;
2. the middle part is stated in terms of the underlying concepts X1, X2, Y1, and Y2; and
3. the unobserved Y1 and Y2 are measured by the observed y1, y2, ...

Fitting the model for different groups of the data

We can fit this model for different groups (say, age groups) by specifying the `group(`*varname*`)` option:

```
(Y1->...) (Y2->-...)                  ///   part 3
(...)                                 ///
(...)                                 ///   part 2
(...)                                 ///
(X1->...) (X2->...),                  ///   part 1
                     group(agegrp)
```

where agegrp is a variable in our dataset, perhaps taking on values 1, 2, 3, ... We can specify the model using the command language or by drawing the model in the GUI and then choosing and filling in the group() option.

After estimation, you can use estat ginvariant (see [SEM] **estat ginvariant**) to obtain Wald tests of whether constraints should be added and score tests of whether constraints should be relaxed.

Which parameters vary by default, and which do not

When we specify `group(`*groupvar*`)`, the measurement parts of the model—parts 1 and 3—are constrained by default to be the same across the groups, whereas the middle part—part 2—will have separate parameters for each group. More specifically, parts 1 and 3 are constrained to be equal across groups except that the variances of the errors will be estimated separately for each group.

If there is no measurement component to the model—if there are no latent variables—then by default all parameters are estimated separately for each group.

Specifying which parameters are allowed to vary in broad, sweeping terms

You can control which parameters are constrained to be equal across groups by specifying the ginvariant() option:

```
(Y1->...) (Y2->...)                    ///   part 3
(...)                                  ///
(...)                                  ///   part 2
(...)                                  ///
(X1->...) (X2->...),                   ///   part 1
                     group(agegrp) ginvariant(classes)
```

The classes are

Class description	Class name	
1. structural coefficients	scoef	
2. structural intercepts	scons	
3. measurement coefficients	mcoef	
4. measurement intercepts	mcons	
5. covariances of structural errors	serrvar	
6. covariances of measurement errors	merrvar	
7. covariances between structural and measurement errors	smerrcov	
8. means of exogenous variables	meanex	(*)
9. covariances of exogenous variables	covex	(*)
10. all of the above	all	(*)
11. none of the above	none	

(*) Be aware that 8, 9, and 10 (meanex, covex, and all) exclude the observed exogenous variables—include only the latent exogenous variables—unless you specify the noxconditional option or the noxconditional option is otherwise implied; see [SEM] **sem option noxconditional**. This is what you would desire in most cases.

The default when ginvariant() is not specified is ginvariant(mcoef mcons):

```
(Y1->...) (Y2->...)                    ///   part 3, measurement
(...)                                  ///
(...)                                  ///   part 2, structural
(...)                                  ///
(X1->...) (X2->...),                   ///   part 1, measurement
                     group(agegrp) ginvariant(mcoef mcons)
```

If you also wanted covariances of errors associated with measurement to be constrained across groups, you could type

```
(Y1->...) (Y2->...)                    ///   part 3, measurement
(...)                                  ///
(...)                                  ///   part 2, structural
(...)                                  ///
(X1->...) (X2->...),                   ///   part 1, measurement
                     group(agegrp) ginvariant(mcoef mcons merrvar)
```

Adding constraints for path coefficients across groups

The ginvariant() option allows you to state in sweeping terms which parameters vary and which are invariant across groups. You can also constrain individual parameters to be equal across groups.

Pretend that in the substantive part of the generic model we have Y1<-Y2. Assume that we fit the model and allow the structural part to vary across groups:

```
(Y1->...) (Y2->...)                     ///    part 3, measurement
(...)                                   ///
(Y1<-Y2)                                ///    part 2, structural
(...)                                   ///
(X1->...) (X2->...),                    ///    part 1, measurement
                  group(agegrp)
```

In this model, the Y1<-Y2 path coefficient is allowed to vary across groups by default. We could constrain the coefficient to be equal across groups by typing

```
(Y1->...) (Y2->...)                     ///    part 3, measurement
(...)                                   ///
(Y1<-Y2@b)                              ///    part 2, structural
(...)                                   ///
(X1->...) (X2->...),                    ///    part 1, measurement
                  group(agegrp)
```

We previously typed (Y1<-Y2). We now type (Y1<-Y2@b). Note the @b.

Constraining a coefficient to equal a symbolic name such as b is how we usually constrain equality, but in the usual case, the symbolic name appears at least twice in our model. For instance, we might have (Y1<-Y2@b) and (Y1<-Y3@b) and thus constrain path coefficients to be equal.

In the case above, however, @b appears only once. Because we specified group(agegrp), results are as if we specified this model separately for each age group, and in each group, we are specifying @b. Thus we are constraining the path coefficient to be equal across all groups.

Adding constraints for means, variances, or covariances across groups

You use the same technique for adding constraints to means, variances, and covariances as you would for adding constraints to path coefficients. Remember that means are specified by the means() option, variances by the variance() option, and covariances by the covariance() option. The variance() and covariance() options are abbreviated var() and cov(), respectively.

You can specify, for instance,

```
(Y1->...) (Y2->...)                     ///    part 3, measurement
(...)                                   ///
(Y1<-Y2)                                ///    part 2, structural
(...)                                   ///
(X1->...) (X2->...),                    ///    part 1, measurement
                  group(agegrp)         ///
                  means(X1@M)
```

to constrain the mean of X1 to be the same across groups. The means would have been different across groups by default.

You can specify

```
(Y1->...) (Y2->...)                     ///   part 3, measurement
(...)                                   ///
(Y1<-Y2)                                ///   part 2, structural
(...)                                   ///
(X1->...) (X2->...),                    ///   part 1, measurement
                    group(agegrp)       ///
                    var(e.Y1@V)
```

to constrain the variance of the error of Y1 to be the same across groups.

If we wanted to allow the errors of Y1 and Y2 to be correlated—by default, errors are uncorrelated—we could add the cov() option:

```
(Y1->...) (Y2->...)                     ///   part 3, measurement
(...)                                   ///
(...)                                   ///   part 2, structural
(...)                                   ///
(X1->...) (X2->...),                    ///   part 1, measurement
                    group(agegrp)       ///
                    var(e.Y1@V)         ///
                    cov(e.Y1*e.Y2)
```

If we then wanted to constrain the covariance to be the same across groups, we would type

```
(Y1->...) (Y2->...)                     ///   part 3, measurement
(...)                                   ///
(...)                                   ///   part 2, structural
(...)                                   ///
(X1->...) (X2->...),                    ///   part 1, measurement
                    group(agegrp)       ///
                    var(e.Y1@V)         ///
                    cov(e.Y1*e.Y2@C)
```

Adding constraints for some groups but not others

Consider the following model:

```
... (Y1<-Y2) ..., group(agegrp)
```

Above we saw how to constrain the Y1<-Y2 path coefficient to be the same across groups:

```
... (Y1<-Y2@b) ..., group(agegrp)
```

To constrain the path coefficients Y1<-Y2 to be equal for groups 1 and 2, but to leave the Y1<-Y2 path coefficients unconstrained for the remaining groups, we could type

```
... (Y1<-Y2) (1: Y1<-Y2@b) (2: Y1<-Y2@b) ..., group(agegrp)
```

Think of this as follows:

1. (Y1<-Y2): We set a path for all the groups.
2. (1: Y1<-Y2@b): We modify the path for agegrp = 1.
3. (2: Y1<-Y2@b): We modify the path for agegrp = 2.
4. We do not modify the path for any other agegrp value.

The result is that we constrain age groups 1 and 2 to have the same value of the path, and we do not constrain the path for the other age groups.

You can constrain variance and covariance estimates to be the same across some groups but not others in the same way. You can specify, for instance,

```
..., group(agegrp) var(1: e.Y1@V) var(2: e.Y1@V)
```

or

```
..., group(agegrp) cov(e.Y1*e.Y2) cov(1: e.Y1*e.Y2@C) ///
                                    cov(2: e.Y1*e.Y2@C)
```

Similarly, you can constrain means for some groups but not others, although this is rarely done:

```
..., group(agegrp) means(1: X1@b) means(2: X1@b)
```

Adding paths for some groups but not others

In the same way that you can constrain coefficients for some groups but not others, you can add paths for some groups but not others. Consider the following model:

```
... (Y1<-Y2) ..., group(agegrp)
```

You can add the path `Y1<-Y3` for groups 1 and 2 by typing

```
... (Y1<-Y2) (1: Y1<-Y3) (2: Y1<-Y3) ..., group(agegrp)
```

You can add covariances for some groups but not others in the same way. For instance, to allow the errors of `Y1` and `Y2` to be correlated in groups 1 and 2 only, you can specify

```
..., group(agegrp) cov(1: e.Y1*e.Y2) cov(2: e.Y1*e.Y2)
```

Relaxing constraints

Just as you can specify

```
..., group(agegrp) ginvariant(classes)
```

and then add constraints, you can also specify

```
..., group(agegrp) ginvariant(classes)
```

and then relax constraints that the classes impose.

For instance, if we specified `ginvariant(scoef)`, then we would be constraining `(Y1<-Y2)` to be invariant across groups. We could then relax that constraint by typing

```
... (Y1<-Y2) (1: Y1<-Y2@b1) (2: Y1<-Y2@b2) ..., ///
                              group(agegrp) ginvariant(scoef)
```

The path coefficients would be free in groups 1 and 2 and constrained in the remaining groups, if there are any. The path coefficient is free in group 1 because we specified symbolic name `b1`, and `b1` appears nowhere else in the model. The path coefficient is free in group 2 because symbolic name `b2` appears nowhere else in the model. If there are remaining groups and we want to relax the constraint on them, too, we would need to add `(3: Y1<-Y2@b3)`, and so on.

The same technique can be used to relax constraints on means, variances, and covariances:

```
..., group(agegrp) ginvariant(... meanex ...)  ///
                   means(1: X1@b1) means(2: X1@b2)

..., group(agegrp) ginvariant(... serrvar ...) ///
                   var(1: e.Y1@V1) var(2: e.Y1@V2)

..., group(agegrp) ginvariant(... svar ...)    ///
                   cov(1: e.Y1*e.Y2@C) cov(2: e.Y1*e.Y2@C)
```

Also see

[SEM] **intro 4** — Tour of models

[SEM] **intro 6** — Postestimation tests and predictions

[SEM] **sem group options** — Fitting models on different groups

[SEM] **sem option covstructure( )** — Specifying covariance restrictions

Title

intro 6 — Postestimation tests and predictions

Description

After fitting a model using sem, you can perform statistical tests, obtain predicted values, and more. Everything you can do is listed below.

Remarks

Remarks are presented under the following headings:

Replaying the model

After estimation, you can type sem without arguments and the output will be redisplayed:

```
. sem
```
(original output reappears)

If you wish to see results in the Bentler–Weeks formulation, after estimation type

```
. estat framework
```
(output omitted)

See [SEM] **example 11**.

In many of the postestimation commands listed below, you will need to refer symbolically to particular coefficients. For instance, in the model

```
. sem ... (Y<-x1) ..., ... cov(e.Y1*e.Y2)
```

the symbolic name of the coefficient corresponding to the path Y<-x1 is _b[Y1:x1], and the symbolic name of the coefficient corresponding to the covariance of e.Y1 and e.Y2 is _b[cov(e.Y1,e.Y2):_cons].

Figuring out what the names are can be difficult, so instead, type

```
. sem, coeflegend
```

sem will produce a table looking very much like the estimation output that lists the _b[] notation for the estimated parameters in the model; see [SEM] **example 8**.

Obtaining goodness-of-fit statistics

One goodness-of-fit statistic and test is reported at the bottom of the sem output:

```
. sem                  // redisplay results
Variables in structural equation model
  (output omitted )
Structural equation model
(coefficient table omitted)
LR test of model vs. saturated: chi2(2)   =      1.78, Prob > chi2 = 0.4111
```

This test is a goodness-of-fit test in badness of fit units; a significant result implies that there may be missing paths in the model's specification. More mathematically, the null hypothesis of this test is that the fitted covariance matrix and mean vector of the observed variables are equal to the matrix and vector observed in the population as measured by the sample. Remember, however, the goal is not to maximize the goodness of fit. One must not add paths that are not theoretically meaningful.

In addition, other goodness-of-fit statistics are available:

1. Command `estat gof` reports a variety of goodness-of-fit statistics; see [SEM] **example 4** and see [SEM] **estat gof**.
2. Command `estat eqgof` reports R^2-like goodness-of-fit statistics for each equation separately; see [SEM] **example 3**.
3. Command `estat ggof` reports goodness-of-fit statistics by group when you have estimated using `sem`'s `group()` option; see [SEM] **example 21**.
4. Command `estat residuals` reports the element-by-element differences between the observed and fitted covariance matrix, and the observed and fitted mean vector, optionally in standardized or in normalized units; see [SEM] **example 10**.
5. Command `estat ic` reports the Akaike and Bayesian information criterion statistics; see [R] **estat**.

Performing tests for including omitted paths and relaxing constraints

1. Command `estat mindices` reports χ^2 modification indices and significance values for each omitted path in the model, along with the expected parameter change; see [SEM] **example 5** and [SEM] **example 9**.
2. Command `estat scoretests` performs score tests on each of the linear constraints placed on the paths and covariances; see [SEM] **example 8**.
3. Command `estat ginvariant` is for use when you have estimated using `sem`'s `group()` option; see [SEM] **intro 5**. This command tests whether you can relax constraints that parameters are equal across groups; see [SEM] **example 22**.

Performing tests of model simplification

1. Command `test` reports Wald tests of single or multiple linear constraints. See [SEM] **example 8** and [SEM] **test**.
2. Command `lrtest` reports likelihood-ratio tests of single or multiple linear constraints. See [SEM] **example 10** and [SEM] **lrtest**.

3. Command `estat eqtest` reports an overall Wald test for each equation in the model, the test corresponding to all coefficients in the equation except the intercept being simultaneously zero; see [SEM] **example 13**.

4. Command `estat ginvariant` is for use when you have estimated using `sem`'s `group()` option; see [SEM] **intro 5**. This command tests whether parameters allowed to vary across groups could be constrained; see [SEM] **example 22**.

Displaying other results, statistics, and tests

1. The `estat stdize:` command prefix—used in front of `test`, `testnl`, `lincom`, and `nlcom`—allows you to perform tests on standardized coefficients. See [SEM] **example 16** and [SEM] **estat stdize**.

2. Command `estat teffects` reports total effects of one variable on another and decomposes the total effect into direct and indirect effects. Results may be reported in standardized or unstandardized form. See [SEM] **example 7** and [SEM] **estat teffects**.

3. Command `estat stable` assesses the stability of nonrecursive structural equation systems; see [SEM] **example 7** and [SEM] **estat stable**.

4. Command `estat summarize` reports summary statistics for the observed variables used in the model; see [SEM] **estat summarize**.

5. Command `lincom` reports the value, standard error, significance, and confidence interval for linear combinations of estimated parameters; see [SEM] **lincom**.

6. Command `nlcom` reports the value, standard error, significance, and confidence interval for nonlinear (and linear) combinations of estimated parameters; see [SEM] **nlcom**.

7. Command `estat vce` reports the variance–covariance matrix of the estimated parameters; see [R] **estat**.

Obtaining predicted values

You obtain predicted values with the `predict` command. Below we will write that predictions are the expected values, but be aware that when there are latent variables in your model, predictions are based on predicted scores; the scores can be inconsistent, and thus any prediction based on them can be inconsistent.

Available are

1. `predict` *newvar*`, xb(`*odepvarname*`)` creates new variable *newvar* containing the predicted values for observed endogenous variable *odepvarname*.

   `predict` *stub*`*, xb` creates new variables *stub*1, *stub*2, ... containing the predicted values for all the observed endogenous variables in the model.

   These predicted values are the expected value of the variable given the values of the observed exogenous variables.

2. `predict` *newvar*`, latent(`*Lname*`)` creates new variable *newvar* containing the predicted values of the latent variable *Lname*, whether endogenous or exogenous.

   `predict` *stub*`*, latent` creates new variables *stub*1, *stub*2, ... containing the predicted values for all the latent variables in the model.

   Predicted values of latent variables, also known as predicted factor scores, are the expected value of the variable given the values of the observed variables.

3. `predict` *newvar*`, xblatent(`*Lname*`)` creates new variable *newvar* containing the predicted values for latent endogenous variable *Lname*.

   `predict` *stub*`*, xblatent` creates new variables *stub*`1`, *stub*`2`, ..., containing the predicted values for all the latent endogenous variables in the model.

   `predict` with `xblatent(`*Lname*`)` differs from `latent(`*Lname*`)` in that the factor scores predicted by `latent()` are then used with the linear equation for *Lname* to make the prediction.

4. `predict` *stub*`*, scores` will create a slew of variables, one for each estimated parameter, containing the observation-by-observation values of the first derivative, also known as scores. This command is intended for use by programmers and may only be used after estimation using `method(ml)` or `method(mlmv)`.

See [SEM] **example 14** and [SEM] **predict**.

Accessing saved results

`sem` saves all results in `e()`; see *Saved results* in [SEM] **sem**. To get some idea of what is stored in `e()` after `sem` estimation, type

```
. ereturn list
  (output omitted)
```

You can save estimation results in files or temporarily in memory and do other useful things with them; see [R] **estimates**.

Not stored by `sem` in `e()` are the Bentler–Weeks matrices, but they can be obtained from the `r()` saved results of `estat framework`.

See [SEM] **sem** and [SEM] **estat framework**.

Also see

[SEM] **intro 5** — Comparing groups

[SEM] **intro 7** — Robust and clustered standard errors

Title

intro 7 — Robust and clustered standard errors

Description

sem provides two options to modify how the standard-error calculations are made: vce(robust) and vce(cluster *clustvar*). These standard errors are less efficient than the default standard errors, but they are valid under less restrictive assumptions.

These options are allowed only when default estimation method method(ml) is used, or option method(mlmv) is used. ml stands for maximum likelihood and mlmv stands for maximum likelihood with missing values; see *Assumptions and choice of estimation method* in [SEM] **intro 3** and see [SEM] **sem**.

Also see [SEM] **intro 8**, entitled *Standard errors, the full story*.

Options

vce(*vcetype*) specifies how the VCE, and thus the standard errors, is calculated. VCE stands for variance–covariance matrix of the estimators. The standard errors that sem reports are the square roots of the diagonal elements of the VCE matrix.

vce(oim) is the default. oim stands for observed information matrix (OIM). The information matrix is the matrix of second derivatives, usually of the log-likelihood function. The OIM estimator of the VCE is based on asymptotic maximum-likelihood theory. The VCE obtained in this way is valid if the errors are independent and identically distributed normal, although the estimated VCE is known to be reasonably robust to violations of the normality assumption, at least as long as the distribution is symmetric and normal-like.

vce(robust) specifies an alternative calculation for the VCE, called robust because the VCE calculated in this way is valid under relaxed assumptions. The method is formally known as the Huber/White/sandwich estimator. The VCE obtained in this way is valid if the errors are independently distributed. It is not required that the errors follow a normal distribution, nor is it required that they be identically distributed from one observation to the next. Thus the vce(robust) VCE is robust to heteroskedasticity of the errors.

vce(cluster *clustvar*) is a generalization of the vce(robust) calculation that relaxes the assumption of independence of the errors and replaces it with the assumption of independence between clusters. Thus the errors are allowed to be correlated within clusters.

Remarks

The vce(robust) option,

```
. sem ..., ... vce(robust)
```

and the vce(cluster *clustvar*) option,

```
. sem ..., ... vce(cluster clustvar)
```

relax assumptions that are sometimes unreasonable for a given dataset and thus produce more accurate standard errors in those cases. Those assumptions are homoskedasticity of the variances of the errors—`vce(robust)`—and independence of the observations—`vce(cluster` *clustvar*`)`. `vce(cluster` *clustvar*`)` relaxes both assumptions.

Homoskedasticity means the variances of the errors are the same from observation to observation. Homoskedasticity can be unreasonable if, for instance, the error corresponds to a dependent variable of income or socioeconomic status. It would not be unreasonable to instead assume that, in the data, the variance of income or socioeconomic status increases as the mean increases. In such cases, rather than typing

```
. sem (y<-...) (...) (...<-x1) (...<-x2)
```

you would type

```
. sem (y<-...) (...) (...<-x1) (...<-x2), vce(robust)
```

Independence means the observations are uncorrelated. If you have observations on people, some of whom live in the same neighborhoods, it would not be unreasonable to assume instead that the error of one person is correlated with those of others who live in the same neighborhood because neighborhoods tend to be homogeneous. In such cases, if you knew the neighborhood, rather than typing

```
. sem (y<-...) (...) (...<-x1) (...<-x2)
```

you would type

```
. sem (y<-...) (...) (...<-x1) (...<-x2), vce(cluster neighborhood)
```

Understand that if the assumptions of independent and identically distributed normal errors are met, the `vce(robust)` and `vce(cluster` *clustvar*`)` standard errors are less efficient than the standard `vce(oim)` standard errors. Less efficient means that for a given sample size, the standard errors jump around more from sample to sample than would the `vce(oim)` standard errors. `vce(oim)` standard errors are unambiguously best when the standard assumptions of homoskedasticity and independence are met.

Also see

[SEM] **intro 6** — Postestimation tests and predictions

[SEM] **intro 8** — Standard errors, the full story

[SEM] **sem option method()** — Specifying method and calculation of VCE

Title

intro 8 — Standard errors, the full story

Description

In [SEM] **intro 7**, we told you part of the story of the calculation of the VCE, the part we wanted to emphasize. In this section, we tell you the full story.

`sem` provides three or four methods for solving the point estimates, depending on how you count, and seven techniques for obtaining the corresponding VCE. They are

Method	Allowed techniques	Comment
ML	OIM	default
	EIM	
	OPG	
	robust	a.k.a. QML
	clustered	
	bootstrap	
	jackknife	
MLMV	OIM	default
	EIM	
	OPG	
	robust	a.k.a. QML
	clustered	
	bootstrap	
	jackknife	
ADF	OIM	robust-like
	EIM	
	bootstrap	
	jackknife	

Abbreviations are

Methods

ML	maximum likelihood
QML	quasimaximum likelihood
MLMV	maximum likelihood with missing values
ADF	asymptotic distribution free

Techniques

OIM	observed information matrix
EIM	expected information matrix
OPG	outer product of the gradients
robust	Huber/White/sandwich estimator
clustered	generalized Huber/White/sandwich estimator
bootstrap	nonparametric bootstrap
jackknife	delete-one jackknife

In *Assumptions and choice of estimation method* of [SEM] **intro 3**, we gave reasons for why you might want to choose methods ML, QML, MLMV, and ADF.

In [SEM] **intro 7**, we gave reasons for why you might want choose techniques OIM, robust, and clustered. EIM has similar properties to OIM and is used in performing score tests. The `sem` command secretly calculates the EIM when necessary so that you can use postestimation score-test commands even if you estimate using a technique other than EIM. EIM is available to you because the `sem` command needs EIM for its own hidden purposes.

By the way, OIM refers to the observed information matrix; it is the inverse of the negative of the matrix of second derivatives. EIM refers to the expected information matrix; it is the inverse of the negative of the expected value of the matrix of second derivatives.

For a discussion of bootstrap and jackknife variance estimation, see [R] **bootstrap** and [R] **jackknife**.

Options

In terms of `sem` options, the table above reads as follows:

method()	vce()	Comment
`method(ml)`	`vce(oim)`	default
	`vce(eim)`	
	`vce(opg)`	
	`vce(robust)`	a.k.a. QML
	`vce(cluster` *clustvar*`)`	
	`vce(bootstrap)`	
	`vce(jackknife)`	
`method(mlmv)`	`vce(oim)`	default
	`vce(eim)`	
	`vce(opg)`	
	`vce(robust)`	a.k.a. QML
	`vce(cluster` *clustvar*`)`	
	`vce(bootstrap)`	
	`vce(jackknife)`	
`method(adf)`	`vce(oim)`	default; `vce(robust)`-like
	`vce(eim)`	
	`vce(bootstrap)`	
	`vce(jackknife)`	

`method(`*emethod*`)` specifies the estimation method `sem` is to use. If `method()` is not specified, then `method(ml)` is assumed.

`vce(`*vcetype*`)` specifies the technique to be used to obtain the VCE. When `vce()` is not specified, then `vce(oim)` is assumed.

Remarks

Although we provide a clean separation between the method used to obtain point estimates and the technique used to obtain the VCE, the literature often does not. For instance, the term QML (quasimaximum likelihood) refers to `method(ml)` or `method(mlmv)` and `vce(robust)`.

In the case of method(adf), even though the default technique is vce(oim), the assumptions justifying the calculations are the same as for vce(robust), namely, that the errors be independent. Thus method(adf) and the default vce(oim) can be used even when errors are heteroskedastic.

Also see

[SEM] **intro 7** — Robust and clustered standard errors

[SEM] **intro 9** — Fitting models using survey data

[SEM] **sem option method()** — Specifying method and calculation of VCE

Title

intro 9 — Fitting models using survey data

Description

Sometimes the data are not a simple random sample from the underlying population but instead are based on a complex survey design that can include stages of clustered sampling and stratification. Estimates produced by `sem` can be adjusted for these issues.

Remarks

Data obtained from surveys, properly treated, produce different point estimates because some observations represent a greater proportion of the underlying population than others and produce different standard errors because the observation-to-observation (sample-to-sample) variation is a function of the survey's design.

To obtain survey-corrected results, you first describe the characteristics of the survey with `svyset`:

```
. svyset county [pw=samplewgt], fpc(n_counties) strata(states) || ///
                                school, fpc(n_schools)          || ///
                                student, fpc(n_students)
```

In the above, we are telling Stata that our data are from a three-stage sampling design. The first stage samples without replacement counties within state; the second, schools within each sampled county; and the third, students within schools.

Once we have done that, we can tell Stata to make the survey adjustment by prefixing statistical commands with `svy:`

```
. svy: regress test_result teachers_per_student sex ...
```

Point estimates and standard errors will be adjusted.

You can use the `svy:` prefix with `sem`

```
. svy: sem (test_result<-...) ... (teachers_per_student->...) ...
```

See the *Stata Survey Data Reference Manual* for more information on this. From a survey perspective, `sem` is not different from any other statistical command of Stata.

Once results are estimated, you do not include the `svy:` prefix in front of the postestimation commands. You type, for instance,

```
. estat eqtest ...
```

You do not type `svy: estat eqtest ....`

Some postestimation procedures you might ordinarily perform can be inappropriate with survey estimation results. This is because you no longer have a sample likelihood value. The postestimation command `lrtest` is an example. If you attempt to use an inappropriate postestimation command, you will be warned.

```
. lrtest ...
lrtest is not appropriate with survey estimation results
r(322);
```

Also see

[SEM] **intro 8** — Standard errors, the full story

[SEM] **intro 10** — Fitting models using summary statistics data

[SVY] *Stata Survey Data Reference Manual*

Title

intro 10 — Fitting models using summary statistics data

Description

In textbooks and research papers, the data used are often printed in summary statistic form. These summary statistics include means, standard deviations or variances, and correlations or covariances. These summary statistics can be used in place of the underlying raw data to fit models using `sem`.

Summary statistics data (SSD) are convenient for publication because of their terseness. By not revealing individual responses, they do not violate participant confidentiality, which is sometimes important.

Remarks

Remarks are presented under the following headings:

Background

The SEM estimator is a function of the first and second moments—the means and covariances—of the data. Thus it is possible to obtain estimates of the parameters of an SEM model using means and covariances. One does not need the original dataset.

In terms of `sem`, one can create a dataset containing these summary statistics and then use that dataset to obtain fitted models. The `sem` command is used just as one would use it with the original, raw data.

How to use sem with SSD

To use `sem` with SSD,

1. You enter the summary statistics using the `ssd` command. How you do that is the topic of an upcoming section.
2. You save the data just as you would any dataset, namely, with the `save` command.
3. You use the `sem` command just as you would ordinarily. You `use` the SSD if they are not already in memory, and no special syntax or options are required by `sem`, except
   a. Do not use `sem`'s `if` *exp* or `in` *range* modifiers. You do not have the raw data in memory and so you cannot select subsets of the data.
   b. If you have entered summary statistics for groups of observations (for example, males and, separately, females), use `sem`'s `select()` option if you want to fit the model using a subset of the groups. That is, where you would ordinarily type

```
. sem ... if sex==1, ...
```

you instead type

```
. sem ..., ... select(1)
```

Where you would ordinarily type

```
. sem ... if region==1 | region==3, ...
```

you instead type

```
. sem ..., ... select(1 3)
```

See [SEM] **example 3**.

What you cannot do with SSD

With SSD in memory,

1. You cannot obtain robust standard errors, which you would normally do by specifying `sem` option `vce(robust)`.
2. You cannot obtain clustered standard errors, which you would normally do by specifying `sem` option `vce(cluster` *clustvar*`)`.
3. You cannot obtain survey-adjusted results, which you would normally do by specifying the `svy:` prefix in front of the `sem` command.
4. You cannot obtain bootstrap or jackknife standard errors, which you would normally do by specifying `sem` option `vce(bootstrap)` or `vce(jackknife)`.
5. You cannot obtain VCE estimates from the observation-level outer product of the gradients, which you would normally do by specifying `vce(opg)`.
6. You cannot use weights, which you would normally do by specifying, for instance, `[fw=`*varname*`]`.
7. You cannot restrict the estimation sample using `if` *exp* or `in` *range*.
8. You cannot fit the model using maximum likelihood with missing values or the asymptotic distribution free method, which you would normally do by specifying `method(mlmv)` or `method(adf)`.

Entering SSD

Entering SSD is easy. You need to see an example of how easy it is before continuing: see [SEM] **example 2**.

What follows is an outline of the procedure. Let us begin with the data you need to have. You have

1. The names of the variables. We will just call them `x1`, `x2`, and `x3`.
2. The number of observations, say, 74.
3. The correlations, say,

```
 1
-0.8072    1
 0.3934   -0.5928    1
```

or you may have the covariances,

```
33.4722
-3.6294    0.6043
 1.0374   -0.2120    0.2118
```

4. The variances, 33.4722, 0.6043, and 0.2118.

   Or the standard deviations, 5.7855, 0.7774, and 0.4602.

   Or neither.

   If you have the covariances in step 3, you in fact have the variances—they are just the diagonal elements of the covariance matrix—but the software will not make you enter the values twice.

5. The means, 21.2973, 3.0195, and 0.2973.

   Or not.

With that information at hand, do the following:

1. Start with no data in memory:

```
. clear all
```

2. Initialize the SSD by stating the names of the variables:

```
. ssd init x1 x2 x3
```

   The remaining steps can be done in any order.

3. Set the number of observations:

```
. ssd set obs 74
```

4. Set the covariances:

```
. ssd set cov 33.4722 \ -3.6294 .6043 \ 1.0374 -.2120 .2118
```

   Or the correlations:

```
. ssd set cor  1 \ -.8072 1 \ .3934 -.5928 1
```

5. If you set covariances in step 4, skip to step 6. Otherwise, if you have them, set the variances:

```
. ssd set var 33.4722 .6043 .2118
```

   Or set the standard deviations:

```
. ssd set sd  5.6855 .7774 .4602
```

6. Set the means if you have them:

```
. ssd set means 21.2973 3.0195 .2973
```

7. If at any point you become confused as to what you have set and what remains to be set, type

```
. ssd status
```

8. If you want to review what you have set, type

```
. ssd list
```

9. If you make a mistake, you can repeat any ssd set command by adding the replace option to the end. For instance, you could reenter the means by typing

```
. ssd set means 21.2973 3.0195 .2973, replace
```

10. Save the dataset just as you would with any dataset:

```
. save mydata
```

You are now ready to use sem with the SSD. With the SSD in memory, you issue the sem command just as you would if you had the raw data,

```
. sem ...
```

Entering SSD for multiple groups

You can enter summary statistics for groups of the data. Perhaps you have summary statistics not for the data as a whole, but for males and for females, or for the young, for the middle-aged, and for the old.

Let's pretend you have the following data:

The young:			
observations:	74		
correlations:	1		
	−0.8072	1	
	0.3934	−0.5928	1
standard deviations:	5.6855	0.7774	0.4602
means:	21.2973	3.0195	0.2973
The middle-aged:			
observations:	141		
correlations:	1		
	−0.5721	1	
	0.3843	−0.4848	1
standard deviations:	4.9112	0.7010	0.5420
means:	38.1512	5.2210	0.2282
The old:			
observations:	36		
correlations:	1		
	−0.8222	1	
	0.3712	−0.3113	1
standard deviations:	6.7827	0.7221	0.4305
means:	58.7171	2.1511	0.1623

The commands for entering these summary statistics are

```
. ssd init x1 x2 x3
. ssd set obs   74
. ssd set cor    1     \  -.8072  1     \ .3934 -.5928 1
. ssd set sd     5.6855   .7774  .4602
. ssd set means 21.2973 3.0195  .2973

. ssd addgroup agecategory
. ssd set obs  141
. ssd set cor    1     \  -.5721  1     \ .3843 -.4848 1
. ssd set sd     4.9112   .7010  .5420
. ssd set means 38.1512 5.2210  .2282

. ssd addgroup
. ssd set obs   36
```

```
. ssd set cor    1    \ -.8222  1    \ .3712 -.3113 1
. ssd set sd       6.7827   .7221   .4305
. ssd set means 58.7171  2.1511   .1623

. save mygroupdata
```

The general procedure is

1. Enter the summary statistics for the first group just as outlined in the previous section.
2. Next add a group by typing

   ```
   . ssd addgroup newgroupvar
   ```

   In that one command, you are telling `ssd` two things. You are telling `ssd` that the summary statistics you entered in step 1 were for a group you are now calling *newgroupvar*, and in particular they were for *newgroupvar* = 1. You are also telling `ssd` that you now want to enter the summary statistics for the next group, namely, *newgroupvar* = 2.
3. Enter the summary statistics for the second group in the same way you entered them for the first group, just as outlined in the previous section.
4. If you have a third group, add it by typing

   ```
   . ssd addgroup
   ```

   In this case, you are telling `ssd` only one thing: that you now want to enter data for the next group, namely, *newgroupvar* = 3.
5. Enter the summary statistics for the third group in the same way you entered them for the second group, and just as outlined in the previous section.
6. If you want to add more groups, continue in the same way. Declare the next group of data by typing

   ```
   . ssd addgroup
   ```

   and then enter the data by using the `ssd set` command.
7. If you mistakenly add a group and wish to rescind that, type

   ```
   . ssd unaddgroup
   ```

8. If you wish to go back and modify the values entered for a previous group, put the group number between `ssd set` and what is being set—for instance, type `ssd set 2 observations` ...—and specify the `replace` option. For instance, to reenter the correlations for group 1, type

   ```
   . ssd set 1 correlations values, replace
   ```

What happens when you do not set all the summary statistics

You are required to set the number of observations and to set the covariances or the correlations. Setting the variances (standard deviations) and setting the means are optional.

1. If you set correlations only, then
   a. Means are assumed to be 0.
   b. Standard deviations are assumed to be 1.
   c. You will not be able to pool across groups if you have group data.

As a result of (a) and (b), the parameters `sem` estimates will be standardized even when you do not specify `sem`'s `standardized` reporting option. Estimated means and intercepts will be zero.

Concerning (c), we need to explain. This concerns group data. If you type

```
. sem ...
```

then `sem` fits a model using all the data. `sem` does that whether you have raw data or SSD in memory. If you have SSD with groups—say, males and females or age groups 1, 2, and 3—`sem` combines the summary statistics to obtain the summary statistics for the overall data. It is only possible to do this when covariances and means are known for each group. If you set correlations without variances or standard deviations and without means, the necessary statistics are not known and the groups cannot be combined. Thus if you type

```
. sem ...
```

you will get an error message. You can still estimate using `sem`; you just have to specify on which group you wish to run, and you do that with the `select()` option:

```
. sem ..., select(#)
```

2. If you set correlations and means,

   a. Standard deviations are assumed to be 1.

   b. You will not be able to pool across groups if you have group data.

   This situation is nearly identical to situation 1. The only difference is that estimated means and intercepts will be nonzero.

3. If you set correlations and standard deviations or variances, or if you set covariances only,

   a. Means are assumed to be 0.

   b. You will not be able to pool across groups

   This situation is a little better than situation 1. Estimated intercepts will be zero, but the remaining estimated coefficients will not be standardized unless you specify `sem`'s `standardized` reporting option.

Labeling SSD

You may use the following commands on SSD, and you use them in the same way you would with an ordinary dataset:

1. `rename` *oldvarname newvarname*
   That is, you may rename the variables; see [D] **rename**.

2. `label data "`*dataset label*`"`
   You may label the dataset; see [D] **label**.

3. `label variable` *varname* `"`*variable label*`"`
   You may label variables; see [D] **label**.

4. `label values` *groupvarname valuelabelname*
   You may place a value label on the group variable; see [D] **label**. The group variable always takes on the values 1, 2, ...

5. `note:` *my note*
   `note` *varname*`:` *my note*
   You may places notes on the dataset or on its variables; see [D] **notes**.

Do not modify the SSD except by using the ssd command. Most importantly, do not drop variables or observations.

Making summary statistics from data for use by others

If you have raw data and wish to make the summary statistics available for subsequent publication, type

```
. ssd build varlist
```

where *varlist* lists the variables you wish to include in the dataset. The SSD will replace the raw data you had in memory. The full syntax is

```
. ssd build varlist if exp in range
```

so you may specify `if` and `in` to restrict the observations that are included.

For instance, to build an SSD for variables `occ_prestige`, `income`, and `social_status`, type

```
. ssd build occ_prestige income social_status
```

If you wish to build the dataset to include separate groups for males and females, type

```
. ssd build occ_prestige income social_status, group(sex)
```

However the `sex` variable was coded in your original data, the two sexes will be now be coded 1 and 2 in the resulting SSD. Which sex is 1 and which is 2 will correspond to however `sort` would have ordered `sex` in its original coding. For instance, if variable `sex` took on values "`male`" and "`female`", the resulting variable `sex` would take on values 1 corresponding to female and 2 corresponding to male.

Once you have built the SSD, you can describe it and list it:

```
. ssd describe
. ssd list
```

See [SEM] **example 25**.

Also see

[SEM] **intro 9** — Fitting models using survey data

[SEM] **ssd** — Making summary statistics data

[SEM] **sem option select()** — Using sem with summary statistics data

[SEM] **example 2** — Creating a dataset from published covariances

[SEM] **example 3** — Two-factor measurement model

[SEM] **example 19** — Creating multiple-group summary statistics data

[SEM] **example 25** — Creating summary statistics data from raw data

Title

estat eqgof — Equation-level goodness-of-fit statistics

Syntax

```
estat eqgof [, format(%fmt)]
```

Menu

Statistics > Structural equation modeling (SEM) > Goodness of fit > Equation-level goodness of fit

Description

estat eqgof displays equation-by-equation goodness-of-fit statistics. Displayed are R^2 and the Bentler–Raykov squared multiple-correlation coefficient (Bentler and Raykov 2000).

These two concepts of fit are equivalent for recursive structural equation models and univariate linear regression. For nonrecursive structural equation models, these measures are distinct.

Equation-level variance decomposition is also reported, along with the overall model coefficient of determination.

Option

format(%*fmt*) specifies the display format. The default is format(%9.0f).

Remarks

See [SEM] **example 3**.

In rare circumstances, these equation-level goodness-of-fit measures in nonrecursive structural equations have unexpected values. It is possible to obtain negative R^2 and multiple-correlation values.

It is recommended to use the Bentler–Raykov squared multiple correlations as a measure of explained variance for nonrecursive systems that involve endogenous variables with reciprocal causations.

Saved results

estat eqgof saves the following in r():

Scalars	
r(N_groups)	number of groups
r(CD[_#])	overall coefficient of determination (for group #)
Matrices	
r(nobs)	sample size for each group
r(eqfit[_#])	fit statistics (for group #)

Also see

[SEM] **example 3** — Two-factor measurement model

[SEM] **estat gof** — Goodness-of-fit statistics

[SEM] **estat ggof** — Group-level goodness-of-fit statistics

[SEM] **methods and formulas** — Methods and formulas

[SEM] **sem postestimation** — Postestimation tools for sem

Title

estat eqtest — Equation-level test that all coefficients are zero

Syntax

```
estat eqtest [, total]
```

Menu

Statistics > Structural equation modeling (SEM) > Testing and CIs > Equation-level Wald tests

Description

`estat eqtest` displays Wald tests that all coefficients excluding the intercept are zero for each equation in the model.

Option

`total` is for use when estimation was with `sem, group()`. It specifies that the tests be aggregated across the groups.

Remarks

See [SEM] **example 13**.

Saved results

`estat eqtest` saves the following in `r()`:

Scalars

`r(N_groups)`	number of groups

Matrices

`r(nobs)`	sample size for each group
`r(test[_#])`	test statistics (for group #)
`r(test_total)`	aggregated test statistics (`total` only)

Also see

[SEM] **example 13** — Equation-level Wald test

[SEM] **test** — Wald test of linear hypotheses

[SEM] **lrtest** — Likelihood-ratio test of linear hypothesis

[SEM] **methods and formulas** — Methods and formulas

[SEM] **sem postestimation** — Postestimation tools for sem

Title

estat framework — Display estimation results in modeling framework

Syntax

`estat framework [, options]`

options	Description
`standardized`	report standardized results
`compact`	display matrices in compact form
`fitted`	include fitted means, variances, and covariances
`format(%fmt)`	display format to use

Menu

Statistics > Structural equation modeling (SEM) > Other > Report model framework

Description

`estat framework` is an `sem` postestimation command that displays the estimation results as a series of matrices derived from the Bentler–Weeks form; see Bentler and Weeks (1980).

Options

`standardized` reports results in standardized form.

`compact` displays matrices in compact form. Zero matrices are displayed as a description. Diagonal matrices are shown as a row vector.

`fitted` displays the fitted mean and covariance values.

`format(%fmt)` specifies the display format to be used. The default is `format(%9.0g)`.

Remarks

See [SEM] **example 11**.

❑ Technical note

If `sem`'s `nm1` option was specified when the model was fit, all covariance matrices are calculated using $N - 1$ in the denominator instead of N.

❑

Saved results

`estat framework` saves the following in `r()`:

Scalars

r(N_groups)	number of groups
r(standardized)	indicator of standardized results (+)

Matrices

r(nobs)	sample size for each group
r(Beta[_#])	coefficients of endogenous variables on endogenous variables (for group #)
r(Gamma[_#])	coefficients of endogenous variables on exogenous variables (for group #)
r(alpha[_#])	intercepts (for group #) (*)
r(Psi[_#])	covariances of errors (for group #)
r(Phi[_#])	covariances of exogenous variables (for group #)
r(kappa[_#])	means of exogenous variables (for group #) (*)
r(Sigma[_#])	fitted covariances (for group #)
r(mu[_#])	fitted means (for group #) (*)

(+) If `r(standardized)`=1, the returned matrices contain standardized values.

(*) If there are no estimated means or intercepts in the `sem` model, these matrices are not returned.

Also see

[SEM] **example 11** — estat framework

[SEM] **intro 6** — Postestimation tests and predictions (*Replaying the model*)

[SEM] **intro 6** — Postestimation tests and predictions (*Accessing saved results*)

[SEM] **methods and formulas** — Methods and formulas

[SEM] **sem postestimation** — Postestimation tools for sem

Title

estat ggof — Group-level goodness-of-fit statistics

Syntax

`estat ggof` [, `format(%`*fmt*`)`]

Menu

Statistics > Structural equation modeling (SEM) > Group statistics > Group-level goodness of fit

Description

`estat ggof` is for use after `sem, group()`. It displays, by group, the standardized root mean squared residual (SRMR), the coefficient of determination (CD), and the model versus saturated χ^2 along with its associated degrees of freedom and p-value.

Option

`format(%`*fmt*`)` specifies the display format. The default is `format(%9.3f)`.

Remarks

See [SEM] **example 21**.

`estat ggof` provides group-level goodness-of-fit statistics after estimation by `sem, group()`; see [SEM] **sem group options**.

Saved results

`estat ggof` saves the following in `r()`:

Scalars

`r(N_groups)`	number of groups

Matrices

`r(gfit)`	fit statistics

Also see

[SEM] **example 21** — Group-level goodness of fit

[SEM] **sem group options** — Fitting models on different groups

[SEM] **estat gof** — Goodness-of-fit statistics

[SEM] **estat eqgof** — Equation-level goodness-of-fit statistics

[SEM] **methods and formulas** — Methods and formulas

[SEM] **sem postestimation** — Postestimation tools for sem

Title

estat ginvariant — Tests for invariance of parameters across groups

Syntax

`estat ginvariant` [, *options*]

options	Description
`showpclass(`*classname*`)`	restrict output to parameters in the specified parameter class
`class`	include joint tests for parameter classes
`legend`	include legend describing parameter classes

classname	Description
`scoef`	structural coefficients
`scons`	structural intercepts
`mcoef`	measurement coefficients
`mcons`	measurement intercepts
`serrvar`	covariances of structural errors
`merrvar`	covariances of measurement errors
`smerrcov`	covariances between structural and measurement errors
`meanex`	means of exogenous variables
`covex`	covariances of exogenous variables
`all`	all of the above
`none`	none of the above

Menu

Statistics > Structural equation modeling (SEM) > Group statistics > Test invariance of parameters across groups

Description

`estat ginvariant` is for use after estimation with `sem, group()`; see [SEM] **sem group options**.

`estat ginvariant` performs score tests (Lagrange multiplier tests) and Wald tests of (1) whether parameters constrained to be equal across groups should be relaxed and (2) whether parameters allowed to vary across groups could be constrained.

See Sörbom (1989) and Wooldridge (2010, 421–428).

Options

showpclass(*classname*) displays tests for the classes specified. showpclass(all) is the default.

class displays a table with joint tests for group invariance for each of the nine parameter classes.

legend displays a legend describing the parameter classes. This option may only be used with the class option.

Remarks

See [SEM] **example 22**.

Saved results

estat ginvariant saves the following in r():

Scalars

r(N_groups)	number of groups

Matrices

r(nobs)	sample size for each group
r(test)	Wald and score tests
r(test_pclass)	parameter classes corresponding to r(test)
r(test_class)	joint Wald and score tests for each class

Also see

[SEM] **example 22** — Testing parameter equality across groups

[SEM] **estat mindices** — Modification indices

[SEM] **estat scoretests** — Score tests

[SEM] **methods and formulas** — Methods and formulas

[SEM] **sem postestimation** — Postestimation tools for sem

Title

estat gof — Goodness-of-fit statistics

Syntax

`estat gof` [, *options*]

options	Description
stats(*statlist*)	statistics to be displayed
nodescribe	suppress descriptions of statistics

statlist	Description
chi2	χ^2 tests; the default
rmsea	root mean squared error of approximation
ic	information indices
indices	indices for comparison against baseline
residuals	measures based on residuals
all	all of the above

Menu

Statistics > Structural equation modeling (SEM) > Goodness of fit > Overall goodness of fit

Description

`estat gof` displays a variety of overall goodness-of-fit statistics after estimation by `sem`.

Options

`stats(`*statlist*`)` specifies the statistics to be displayed. The default is `stats(chi2)`.

`stats(chi2)` reports the model versus saturated test and the baseline versus saturated test. The saturated model is the model that fits the covariances perfectly.

The model versus saturated test is a repeat of the test reported at the bottom of the `sem` output.

In the baseline versus saturated test, the baseline model includes the means and variances of all observed variables plus the covariances of all observed exogenous variables. For a covariance model (a model with no endogenous variables), the baseline includes only the means and variances of observed variables. Be aware that different authors define the baseline model differently.

`stats(rmsea)` reports the root mean squared error of approximation (RMSEA) and its 90% confidence interval, and pclose, the p-value for a test of close fit, namely, RMSEA < 0.05. Most interpreters of this test label the fit close if the lower bound of the 90% CI is below 0.05 and label the fit poor if the upper bound is above 0.10. See Browne and Cudeck (1993).

`stats(ic)` reports the Akaike information criterion (AIC) and Bayesian (or Schwarz) information criterion (BIC). These statistics are available only after estimation using `sem method(ml)` or `method(mlmv)`. These statistics are used not to judge fit in absolute terms but instead to compare the fit of different models. Smaller values indicate a better fit. Be aware that there are many variations (minor adjustments) to statistics labeled AIC and BIC. Reported here are statistics that match `estat ic`; see [R] **estat**. To compare models using statistics, such as AIC and BIC, that are based on likelihoods, models should include the same variables; see [SEM] **lrtest**. See Akaike (1987), Schwarz (1978), and Raftery (1993).

`stats(indices)` reports CFI and TLI, two indices such that a value close to 1 indicates a good fit. CFI stands for comparative fit index. TLI stands for Tucker–Lewis index and is also known as the nonnormed fit index. See Bentler (1990).

`stats(residuals)` reports the standardized root mean squared residual (SRMR) and the coefficient of determination (CD).

A perfect fit corresponds to an SRMR of 0. A good fit is a small value, considered by some to be limited to 0.08. SRMR is calculated using the first and second moments unless `sem` option `nomeans` was specified or implied, in which case SRMR is calculated based on second moments only. Some software packages ignore the first moments even when available. See Hancock and Mueller (2006, 157).

Concerning CD, a perfect fit corresponds to a CD of 1. CD is like R^2 for the whole model.

`stats(all)` reports all the statistics. You can also specify just the statistics you wish reported, such as

```
. estat gof, stats(indices residuals)
```

`nodescribe` suppresses the descriptions of the goodness-of-fit measures.

Remarks

See [SEM] **example 4**.

Saved results

estat gof saves the following in r():

Scalars

r(chi2_ms)	test of target model against saturated model
r(df_ms)	degrees of freedom for r(chi2_ms)
r(p_ms)	p-value for r(chi2_ms)
r(chi2_bs)	test of baseline model against saturated model
r(df_bs)	degrees of freedom for r(chi2_bs)
r(p_bs)	p-value for r(chi2_bs)
r(rmsea)	root mean squared error of approximation
r(lb90_rmsea)	lower bound of 90% CI for RMSEA
r(ub90_rmsea)	upper bound of 90% CI for RMSEA
r(pclose)	p-value for test of close fit: RMSEA < 0.05
r(aic)	Akaike information criterion
r(bic)	Bayesian information criterion
r(cfi)	comparative fit index
r(tli)	Tucker–Lewis fit index
r(cd)	coefficient of determination
r(srmr)	standardized root mean squared residual
r(N_groups)	number of groups

Matrices

r(nobs)	sample size for each group

Also see

[SEM] **example 4** — Goodness-of-fit statistics

[SEM] **estat ggof** — Group-level goodness-of-fit statistics

[SEM] **estat eqgof** — Equation-level goodness-of-fit statistics

[SEM] **estat residuals** — Display mean and covariance residuals

[R] **estat** — Postestimation statistics

[SEM] **methods and formulas** — Methods and formulas

[SEM] **sem postestimation** — Postestimation tools for sem

Title

estat mindices — Modification indices

Syntax

`estat mindices` [, *options*]

options	Description
`showpclass(`*classname*`)`	restrict output to parameters in the specified parameter classes
`minchi2(`*#*`)`	display only tests with modification index (MI) ≥ *#*

classname	Description
`scoef`	structural coefficients
`scons`	structural intercepts
`mcoef`	measurement coefficients
`mcons`	measurement intercepts
`serrvar`	covariances of structural errors
`merrvar`	covariances of measurement errors
`smerrcov`	covariances between structural and measurement errors
`meanex`	means of exogenous variables
`covex`	covariances of exogenous variables
`all`	all of the above
`none`	none of the above

Menu

Statistics > Structural equation modeling (SEM) > Testing and CIs > Modification indices

Description

`estat mindices` reports modification indices for omitted paths in the fitted model. Modification indices are score tests (Lagrange multiplier tests) for the statistical significance of the omitted paths. See Sörbom (1989) and Wooldridge (2010, 421–428).

Options

`showpclass(`*classname*`)` specifies that results be limited to parameters that belong to the specified parameter classes. The default is `showpclass(all)`.

minchi2(#) suppresses listing paths with modification indices (MIs) less than #. By default, estat mindices lists values significant at the 0.05 level, corresponding to $\chi^2(1)$ value minchi2(3.8414588). Specify minchi2(0) if you wish to see all tests.

Remarks

See [SEM] **example 5**.

Saved results

estat mindices saves the following in r():

Scalars

r(N_groups)	number of groups

Matrices

r(nobs)	sample size for each group
r(mindices_pclass)	parameter class of modification indices
r(mindices)	matrix containing the displayed table values

Also see

[SEM] **example 5** — Modification indices

[SEM] **estat scoretests** — Score tests

[SEM] **estat ginvariant** — Tests for invariance of parameters across groups

[SEM] **methods and formulas** — Methods and formulas

[SEM] **sem postestimation** — Postestimation tools for sem

Title

estat residuals — Display mean and covariance residuals

Syntax

`estat residuals [, options]`

options	Description
`normalized`	report normalized residuals
`standardized`	report standardized residuals
`sample`	use sample covariances in residual variance calculations
`nm1`	use adjustment $N - 1$ in residual variance calculations
`zerotolerance(`*tol*`)`	apply tolerance to treat residuals as zero
`format(`*%fmt*`)`	display format

Menu

Statistics > Structural equation modeling (SEM) > Goodness of fit > Matrices of residuals

Description

`estat residuals` displays the mean and covariance residuals after estimation by `sem`. Normalized and standardized residuals are available.

Both mean and covariance residuals are reported unless `sem`'s option `nomeans` was specified or implied at the time the model was fit, in which case mean residuals are not reported.

`estat residuals` usually does not work following `sem` models fit using `method(mlmv)`. It also does not work if there are any missing values, which after all, is the whole point of using `method(mlmv)`.

Options

`normalized` and `standardized` are alternatives. If neither is specified, raw residuals are reported.

Normalized residuals and standardized residuals attempt to adjust the residuals in the same way but go about it differently. The normalized residuals are always valid, but they do not follow a standard normal distribution. The standardized residuals do follow a standard normal distribution but only if they can be calculated; otherwise, they will equal missing values. When both can be calculated (equivalent to both being appropriate), the normalized residuals will be a little smaller than the standardized residuals. See Jöreskog and Sörbom (1986).

`sample` specifies that the sample variance and covariances be used in variance formulas to compute normalized and standardized residuals. The default uses fitted variance and covariance values as described by Bollen (1989).

`nm1` specifies that the variances be computed using $N-1$ in the denominator rather than using sample size N.

`zerotolerance(`*tol*`)` treats residuals within *tol* of zero as if they were zero. *tol* must be a numeric value less than one. The default is `zerotolerance(0)`, meaning that no tolerance is applied. When standardized residuals cannot be calculated, it is because a variance calculated by the Hausman (1978) theorem turns negative. Applying a tolerance to the residuals turns some residuals into 0 and then division by the negative variance becomes irrelevant, and that may be enough to solve the calculation problem.

`format(%`*fmt*`)` specifies the display format. The default is `format(%9.3f)`.

Remarks

See [SEM] **example 10**.

Saved results

`estat residuals` saves the following in `r()`:

Scalars

`r(N_groups)`	number of groups

Macros

`r(sample)`	empty or `sample`, if `sample` was specified
`r(nm1)`	empty or `nm1`, if `nm1` was specified

Matrices

`r(nobs)`	sample size for each group
`r(res_mean[_#])`	raw mean residuals (for group #) (*)
`r(res_cov[_#])`	raw covariance residuals (for group #)
`r(nres_mean[_#])`	normalized mean residuals (for group #) (*)
`r(nres_cov[_#])`	normalized covariance residuals (for group #)
`r(sres_mean[_#])`	standardized mean residuals (for group #) (*)
`r(sres_cov[_#])`	standardized covariance residuals (for group #)

(*) If there are no estimated means or intercepts in the `sem` model, these matrices are not returned.

Also see

[SEM] **example 10** — MIMIC model

[SEM] **estat gof** — Goodness-of-fit statistics

[SEM] **estat ggof** — Group-level goodness-of-fit statistics

[SEM] **estat eqgof** — Equation-level goodness-of-fit statistics

[SEM] **methods and formulas** — Methods and formulas

[SEM] **sem postestimation** — Postestimation tools for sem

Title

estat scoretests — Score tests

Syntax

`estat scoretests [ , minchi2(#) ]`

Menu

Statistics > Structural equation modeling (SEM) > Testing and CIs > Score tests of linear constraints

Description

`estat scoretests` displays score tests (Lagrange multiplier tests) for each of the user-specified linear constraints imposed on the model when it was fit. See Sörbom (1989) and Wooldridge (2010, 421–428).

Option

`minchi2(#)` suppresses output of tests with $\chi^2(1) < \#$. By default, `estat mindices` lists values significant at the 0.05 level, corresponding to $\chi^2(1)$ value `minchi2(3.8414588)`. Specify `minchi2(0)` if you wish to see all tests.

Remarks

See [SEM] **example 8**.

Saved results

`estat scoretests` saves the following in `r()`:

Scalars

`r(N_groups)`	number of groups

Matrices

`r(nobs)`	sample size for each group
`r(Cns_sctest)`	matrix containing the displayed table values

Also see

[SEM] **example 8** — Testing that coefficients are equal, and constraining them

[SEM] **estat mindices** — Modification indices

[SEM] **estat ginvariant** — Tests for invariance of parameters across groups

[SEM] **methods and formulas** — Methods and formulas

[SEM] **sem postestimation** — Postestimation tools for sem

Title

estat stable — Check stability of nonrecursive system

Syntax

`estat stable [, detail]`

Menu

Statistics > Structural equation modeling (SEM) > Other > Assess stability of nonrecursive systems

Description

`estat stable` reports the eigenvalue stability index for nonrecursive models after estimation by `sem`. The stability index is computed as the maximum modulus of the eigenvalues for the matrix of coefficients on endogenous variables predicting other endogenous variables. When the model was fit by `sem` with the `group()` option, `estat stable` reports the index for each group separately.

There are two formulas commonly used to calculate the index. `estat stable` uses the formulation of Bentler and Freeman (1983).

Option

`detail` displays the matrix of coefficients on endogenous variables predicting other endogenous variables, also known as the β matrix.

Remarks

See *nonrecursive (structural) model (system)* in [SEM] **Glossary**. The issue of stability is described there. Also see *Remarks* of [SEM] **estat teffects**.

Saved results

`estat stable` saves the following in `r()`:

Scalars

`r(N_groups)`	number of groups
`r(stindex[_#])`	stability index (for group #)

Matrices

`r(nobs)`	sample size for each group
`r(Beta[_#])`	coefficients of endogenous variables on endogenous variables (for group #)
`r(Re[_#])`	real parts of the eigenvalues of A (for group #)
`r(Im[_#])`	imaginary parts of the eigenvalues of A (for group #)
`r(Modulus[_#])`	modulus of the eigenvalues of A (for group #)

Also see

[SEM] **estat teffects** — Decomposition of effects into total, direct, and indirect

[SEM] **methods and formulas** — Methods and formulas

[SEM] **sem postestimation** — Postestimation tools for sem

Title

estat stdize — Test standardized parameters

Syntax

```
estat stdize: test ...

estat stdize: lincom ...

estat stdize: testnl ...

estat stdize: nlcom ...
```

Menu

Statistics > Structural equation modeling (SEM) > Testing and CIs > Testing standardized parameters

Description

`estat stdize:` can be used to prefix `test`, `lincom`, `testnl`, and `nlcom`; see [SEM] **test**, [SEM] **lincom**, [SEM] **testnl**, and [SEM] **nlcom**.

These commands, without a prefix, work in the underlying metric of SEM, which is to say, path coefficients, variances, and covariances. If the commands are prefixed with `estat stdize:`, they will work in the metric of standardized coefficients and correlation coefficients. There is no counterpart to variances in the standardized metric because variances are standardized to be 1.

Remarks

See [SEM] **example 16**.

Exercise caution when using the `estat stdize:` prefix to perform tests on estimated second moments, which is to say, correlations. Do not test that correlations are zero. Instead, omit the `estat stdize:` prefix and test that covariances are zero. Covariances are more likely to be normally distributed than are correlations.

Saved results

Saved results are the results saved by the command being used with the `estat stdize:` prefix.

Also see

[SEM] **example 16** — Correlation

[SEM] **test** — Wald test of linear hypotheses

[SEM] **lincom** — Linear combinations of parameters

[SEM] **testnl** — Wald test of nonlinear hypotheses

[SEM] **nlcom** — Nonlinear combinations of parameters

[SEM] **sem postestimation** — Postestimation tools for sem

[SEM] **methods and formulas** — Methods and formulas

Title

estat summarize — Report summary statistics for estimation sample

Syntax

`estat summarize` [*eqlist*] [, `group` *estat_summ_options*]

Menu

Statistics > Postestimation > Reports and statistics

Description

`estat summarize` reports the summary statistics in the estimation sample for the observed variables in the model. `estat summarize` is a standard postestimation feature described in [R] **estat**.

`estat summarize` is mentioned here because

1. `estat summarize` cannot be used when `sem` was run on summary statistics data; see [SEM] **intro 10**.
2. `estat summarize` allows the additional option `group` after estimation by `sem`.

Options

`group` may be specified if `group(`*varname*`)` was specified with `sem` at the time the model was fit. It requests that summary statistics be reported by group.

estat_summ_options are the standard options allowed by `estat summarize` and are outlined in *Options for estat summarize* of [R] **estat**.

Saved results

See *Saved results* of [R] **estat**.

Also see

[R] **estat** — Postestimation statistics

[SEM] **sem postestimation** — Postestimation tools for sem

Title

estat teffects — Decomposition of effects into total, direct, and indirect

Syntax

estat teffects [, *options*]

options	Description
compact	do not display effects with no path
standardized	report standardized effects
nolabel	display group values, not labels
nodirect	do not display direct effects
noindirect	do not display indirect effects
nototal	do not display total effects
display_options	control column formats, row spacing, and display of omitted paths

Menu

Statistics > Structural equation modeling (SEM) > Testing and CIs > Direct and indirect effects

Description

estat teffects reports direct, indirect, and total effects for each path (Sobel 1987), along with standard errors obtained by the delta method, after estimation with sem.

Options

compact is a popular option. Consider the following model:

```
. sem (y1<-y2 x1) (y2<-x2)
```

x2 has no direct effect on y1 but does have an indirect effect. estat teffects formats all of its effects tables the same way by default, so there will be a row for the direct effect of x2 on y1 just because there is a row for the indirect effect of x2 on y1. The value reported for the direct effect, of course, will be zero. compact says to omit these unnecessary rows.

standardized reports effects in standardized form, but standard errors of the standardized effects are not reported.

nolabel is relevant only when estimation was with sem's group() option and the group variable has a value label. Groups are identified by group value rather than label.

nodirect, noindirect, and nototal suppress the display of the indicated effect. The default is to display all effects.

display_options: noomitted, vsquish, cformat(%*fmt*), pformat(%*fmt*), sformat(%*fmt*), and nolstretch; see [R] **estimation options**. Although estat teffects is not an estimation command, it allows these options.

Remarks

See [SEM] **example 7**.

Direct effects are the path coefficients in the model.

Indirect effects are all mediating effects. For instance, consider

```
. sem ... (y1<-y2) (y1<-x2) (y2<-x3) ..., ...
```

The direct effect of y2 on y1 is the path coefficient (y1<-y2).

In this example, changes in x3 affect y1, too. That is called the indirect effect and is the product of the path coefficients (y2<-x3) and (y1<-y2). If there were other paths in the model such that y1 changed when x3 changed, those effects would be added to the indirect effect as well. estat teffects reports total indirect effects.

The total effect is the sum of the direct and indirect effects.

When feedback loops are present in the model, such as

```
. sem ... (y1<-y2) (y1<-x2) (y2<-x3 y1) ..., ...
```

care must be taken when interpreting indirect effects. The feedback loop is when a variable indirectly affects itself, as y1 does in the example. y1 affects y2 and y2 affects y1. Thus in calculating the indirect effect, the sum has an infinite number of terms, although the term values get smaller and smaller and thus usually converge to a finite result. It is important that you check nonrecursive models for stability; see Bollen (1989, 397) and see [SEM] **estat stable**. Caution: if the model is unstable, the calculation of the indirect effect can sometimes still converge to a finite result.

Saved results

estat teffects saves the following in r():

Scalars

r(N_groups)	number of groups

Matrices

r(nobs)	sample size for each group
r(direct)	direct effects
r(indirect)	indirect effects
r(total)	total effects
r(V_direct)	covariance matrix of the direct effects
r(V_indirect)	covariance matrix of the indirect effects
r(V_total)	covariance matrix of the total effects

estat teffects with the standardized option additionally saves the following in r():

Matrices

r(direct_std)	standardized direct effects
r(indirect_std)	standardized indirect effects
r(total_std)	standardized total effects

Also see

[SEM] **estat stable** — Check stability of nonrecursive system

[SEM] **methods and formulas** — Methods and formulas

[SEM] **sem postestimation** — Postestimation tools for sem

Title

example 1 — Single-factor measurement model

Description

The single-factor measurement model is demonstrated using the following data:

```
. use http://www.stata-press.com/data/r12/sem_1fmm
(single-factor measurement model)

. summarize

    Variable |       Obs        Mean    Std. Dev.       Min        Max
-------------+--------------------------------------------------------
          x1 |       123    96.28455    14.16444         54        131
          x2 |       123    97.28455    16.14764         64        135
          x3 |       123    97.09756    15.10207         62        138
          x4 |       123    690.9837    77.50737        481        885

. notes

_dta:
  1.  fictional data
  2.  Variables x1, x2, and x3 each contain a test score designed to measure X.
      The test is scored to have mean 100.
  3.  Variable x4 is also designed to measure X, but designed to have mean 700.
```

See *Single-factor measurement models* in [SEM] **intro 4** for background.

Remarks

Remarks are presented under the following headings:

Single-factor measurement model
The measurement error model interpretation

Single-factor measurement model

Below we fit the model:

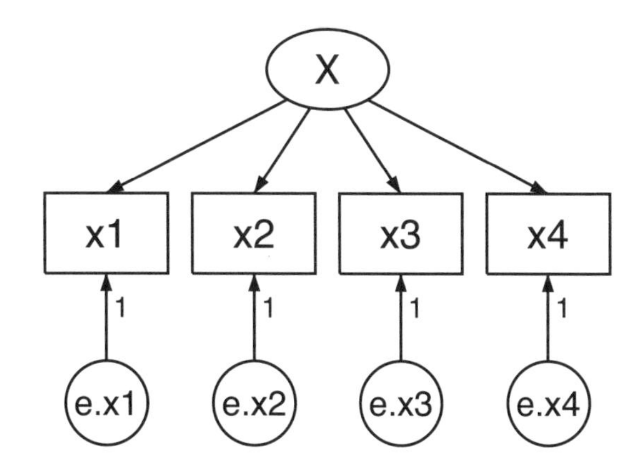

```
. sem (x1 x2 x3 x4 <- X)
Endogenous variables
Measurement:  x1 x2 x3 x4
Exogenous variables
Latent:       X
Fitting target model:
Iteration 0:   log likelihood = -2081.0258
Iteration 1:   log likelihood =  -2080.986
Iteration 2:   log likelihood = -2080.9859
Structural equation model                        Number of obs      =        123
Estimation method  = ml
Log likelihood     = -2080.9859
 ( 1)  [x1]X = 1

------------------------------------------------------------------------------
             |                 OIM
             |      Coef.   Std. Err.      z    P>|z|     [95% Conf. Interval]
-------------+----------------------------------------------------------------
Measurement  |
  x1 <-      |
           X |          1  (constrained)
       _cons |   96.28455   1.271963    75.70   0.000     93.79155    98.77755
  -----------+----------------------------------------------------------------
  x2 <-      |
           X |   1.172364   .1231777     9.52   0.000     .9309398    1.413788
       _cons |   97.28455   1.450053    67.09   0.000      94.4425    100.1266
  -----------+----------------------------------------------------------------
  x3 <-      |
           X |   1.034523   .1160558     8.91   0.000     .8070579    1.261988
       _cons |   97.09756   1.356161    71.60   0.000     94.43953    99.75559
  -----------+----------------------------------------------------------------
  x4 <-      |
           X |   6.886044   .6030898    11.42   0.000     5.704009    8.068078
       _cons |   690.9837   6.960137    99.28   0.000     677.3421    704.6254
-------------+----------------------------------------------------------------
Variance     |
        e.x1 |   80.79361   11.66414                      60.88206    107.2172
        e.x2 |   96.15861   13.93945                      72.37612    127.7559
        e.x3 |   99.70874   14.33299                      75.22708    132.1576
        e.x4 |   353.4711   236.6847                      95.14548    1313.166
           X |   118.2068   23.82631                      79.62878    175.4747
------------------------------------------------------------------------------
LR test of model vs. saturated: chi2(2)   =      1.78, Prob > chi2 = 0.4111
```

The equations for this model are

$$x_1 = \alpha_1 + X\beta_1 + e.x_1$$

$$x_2 = \alpha_2 + X\beta_2 + e.x_2$$

$$x_3 = \alpha_3 + X\beta_3 + e.x_3$$

$$x_4 = \alpha_4 + X\beta_4 + e.x_4$$

Notes:

1. Variable `X` is latent exogenous and thus needs a normalizing constraint. The variable is anchored to the first observed variable, `x1`, and thus the path coefficient is constrained to be 1. See *Identification 2: Normalization constraints (anchoring)* in [SEM] **intro 3**.

2. The path coefficients for X->x1, X->x2, and X->x3 are 1 (constrained), 1.17, and 1.03. Meanwhile, the path coefficient for X->x4 is 6.89. This is not unexpected; we at StataCorp generated this data and the true coefficients are 1, 1, 1, and 7.

3. A test for "model versus saturated" is reported at the bottom of the output; the $\chi^2(2)$ statistic is 1.78 and its significance level is 0.4111. We cannot reject the null hypothesis of this test.

   This test is a goodness-of-fit test in badness-of-fit units; a significant result implies that there may be missing paths in the model's specification.

   More mathematically, the null hypothesis of the test is that the fitted covariance matrix and mean vector of the observed variables are equal to the matrix and vector observed in the population.

The measurement error model interpretation

As we pointed out in *Using path diagrams to specify the model* in [SEM] **intro 2**, if we rename variable x4 to be y, we can reinterpret this measurement model as a measurement *error* model. In this interpretation, X is the unobserved true value. x1, x2, and x3 are each measurements of X, but with error. Meanwhile, y (x4) is really something else entirely. Perhaps y is earnings, and we believe

$$\mathtt{y} = \alpha_4 + \beta_4 \mathtt{X} + \mathtt{e.y}$$

We are interested in β_4, the effect of true X on y.

If we were to go back to the data and type regress y x1, we would obtain an estimate of β_4, but we would expect that estimate to be biased toward zero because of the errors-in-variable problem. The same applies for y on x2 and y on x3. If we do that, we obtain

β_4 based on regress y x1 4.09

β_4 based on regress y x2 3.71

β_4 based on regress y x3 3.70

In the sem output above, we have an estimate of β_4 with the bias washed away:

β_4 based on sem (y<-X) 6.89

The number 6.89 is the value reported for (x4<-X) in the sem output.

That β_4 might be 6.89 seems plausible because we do expect that the estimate should be larger than the estimates we obtain using the variables measured with error. In fact, we can tell you that the 6.89 estimate is quite good because we at StataCorp know that the true value of β_4 is 7. Here is how we manufactured this fictional dataset:

```
set seed 12347
set obs 123
gen X  = round(rnormal(0,10))
gen x1 = round(100 + X + rnormal(0, 10))
gen x2 = round(100 + X + rnormal(0, 10))
gen x3 = round(100 + X + rnormal(0, 10))
gen x4 = round(700 + 7*X + rnormal(0, 10))
```

The data recorded in sem_1fmm.dta was obviously generated using normality, the same assumption that is most often used to justify the SEM ML estimator. In *Assumptions and choice of estimation method* in [SEM] **intro 3**, we explained that the normality assumption can be relaxed and conditional normality can usually be substituted in its place.

So let's consider nonnormal data. Let's make X be $\chi^2(2)$, a violently nonnormal distribution, resulting in the data-manufacturing code

```
set seed 12347
set obs 123
gen X  = (rchi2(2)-2)*(10/2)
gen x1 = round(100 + X + rnormal(0, 10))
gen x2 = round(100 + X + rnormal(0, 10))
gen x3 = round(100 + X + rnormal(0, 10))
gen x4 = round(700 + 7*X + rnormal(0, 10))
```

All the `rnormal()` functions remaining in our code have to do with the assumed normality of the errors. The multiplicative and additive constants in the generation of X simply rescale the $\chi^2(2)$ variable to have mean 100 and standard deviation 10, which would not be important except for the subsequent `round()` functions, which themselves were unnecessary except that we wanted to produce a pretty dataset when we created the original `sem_1fmm.dta`.

In any case, if we rerun the commands using these data, we obtain

β_4 based on `regress y x1`	3.93
β_4 based on `regress y x2`	4.44
β_4 based on `regress y x3`	3.77
β_4 based on `sem (y<-X)`	6.70

We will not burden you with the details of running simulations to assess coverage; we will just tell you that coverage is excellent: reported test statistics and significance levels can be trusted.

By the way, errors in the variables is something that does not go away with larger and larger sample sizes. Change the code above to produce a 100,000-observation dataset instead of a 123-observation one, and you will obtain

β_4 based on `regress y x1`	3.51
β_4 based on `regress y x2`	3.51
β_4 based on `regress y x3`	3.48
β_4 based on `sem (y<-X)`	7.00

Also see

[SEM] **sem** — Structural equation model estimation command

[SEM] **intro 4** — Tour of models

[SEM] **example 2** — Creating a dataset from published covariances

[SEM] **example 3** — Two-factor measurement model

[SEM] **example 24** — Reliability

Title

example 2 — Creating a dataset from published covariances

Description

Williams, Eaves, and Cox (2002) publish covariances from their data. We will use those published covariances in [SEM] **example 3** to fit an SEM model.

In this example, we show how we create the summary statistics dataset (SSD) that we will analyze in that example.

Remarks

Remarks are presented under the following headings:

Background
Creating the SSD
At this point, we could save the dataset and stop
Labeling the SSD
Listing the SSD

For more explanation, also see [SEM] **intro 10**.

Background

In Williams, Eaves, and Cox (2002), the authors report a covariance matrix in a table that looks something like this:

	Affective		Miniscale		Cognitive	Miniscale	
	1	2	...	5	1	2	...
Affective							
1	2038.035	1631.766	...	1721.830	659.795	779.519	...
2		1932.163	...	1688.292	702.969	790.488	...
.							
.							
5				2061.875	775.118	871.211	...
Cognitive							
1					630.518	500.128	...
.							
.							

Creating the SSD

```
. clear all
. ssd init  a1 a2 a3 a4 a5  c1 c2 c3 c4 c5
Summary statistics data initialized.  Next use, in any order,
    ssd set observations (required)
        It is best to do this first.
    ssd set means (optional)
        Default setting is 0.
    ssd set variances or ssd set sd (optional)
        Use this only if you have set or will set correlations and, even
        then, this is optional but highly recommended.  Default setting is 1.
    ssd set covariances or ssd set correlations (required)
. ssd set obs 216
  (value set)
    Status:
                      observations:    set
                             means:  unset
                   variances or sd:  unset
       covariances or correlations:  unset (required to be set)
. #delimit ;
delimiter now ;
. ssd set cov 2038.035 \
>             1631.766 1932.163 \
>             1393.567 1336.871 1313.809 \
>             1657.299 1647.164 1273.261 2034.216 \
>             1721.830 1688.292 1498.401 1677.767 2061.875 \
>              659.795  702.969  585.019  656.527  775.118 630.518 \
>              779.519  790.448  653.734  764.755  871.211 500.128 741.767 \
>
>              899.912  879.179  750.037  897.441 1008.287 648.935 679.970
>             1087.409 \
>
>              775.235  739.157  659.867  751.860  895.616 550.476 603.950
>              677.865  855.272 \
>
>              821.170  785.419  669.951  802.825  877.681 491.042 574.775
>              686.391  622.830  728.674 ;
  (values set)
    Status:
                      observations:    set
                             means:  unset
                   variances or sd:    set
       covariances or correlations:    set
. #delimit cr
delimiter now cr
```

Notes:

1. We used `#delimit` to temporarily set the end-of-line character to semicolon. That was not necessary, but it made it easier to enter the data in a way that would be subsequently more readable. You can use `#delimit` only in do-files; see [P] **#delimit**.
2. We recommend entering SSD using do-files. That way, you can edit the file and get it right.
3. We did not have to reset the delimiter. We could have entered the numbers on one (long) line. That works well when there are only a few summary statistics.

At this point, we could save the dataset and stop

We could save the dataset and stop right now if we wished:

```
. save sem_2fmm
file sem_2fmm.dta saved
```

Obviously, we can save the dataset anytime we wish. We know we could stop because `ssd status` tells us whether there is anything more that we need to define:

```
. ssd status
    Status:
                         observations:    set
                                means:  unset
                      variances or sd:    set
          covariances or correlations:    set
```

Notes:

1. The means have not been set. The authors did not provide the means.
2. `ssd status` would mention if anything that was not set was required to be set.

Labeling the SSD

If we were to use `ssd describe` to describe these data, the output would look like this:

```
. ssd describe
  Summary statistics data
    obs:             216
   vars:              10

  variable name                 variable label

  a1
  a2
  a3
  a4
  a5
  c1
  c2
  c3
  c4
  c5
```

We can add labels and notes to our dataset:

```
. label data "Affective and cognitive arousal"
. label var a1 "affective arousal 1"
. label var a2 "affective arousal 2"
. label var a3 "affective arousal 3"
. label var a4 "affective arousal 4"
. label var a5 "affective arousal 5"
. label var c1 "cognitive arousal 1"
. label var c2 "cognitive arousal 2"
. label var c3 "cognitive arousal 3"
. label var c4 "cognitive arousal 4"
```

```
. label var c5 "cognitive arousal 5"
. #delimit ;
delimiter now ;
. notes: Summary statistics data containing published covariances
>        from Thomas O. Williams, Ronald C. Eaves, and Cynthia Cox,
>        2 Apr 2002, "Confirmatory factor analysis of an instrument
>        designed to measure affective and cognitive arousal",
>        _Educational and Psychological Measurement_,
>        vol. 62 no. 2, 264-283. ;
. notes: a1-a5 report scores from 5 miniscales designed to measure
>        affective arousal. ;
. notes: c1-c5 report scores from 5 miniscales designed to measure
>        cognitive arousal. ;
. notes: The series of tests, known as the VST II
>        (Visual Similes Test II) were administered to 216 children
>        ages 10 to 12.  The miniscales are sums of scores of
>        5 to 6 items in VST II. ;
. #delimit cr
delimiter now cr
. ssd describe

  Summary statistics data
    obs:             216                Affective and cognitive arousal
   vars:              10
                                        (_dta has notes)

  variable name                   variable label

  a1                              affective arousal 1
  a2                              affective arousal 2
  a3                              affective arousal 3
  a4                              affective arousal 4
  a5                              affective arousal 5
  c1                              cognitive arousal 1
  c2                              cognitive arousal 2
  c3                              cognitive arousal 3
  c4                              cognitive arousal 4
  c5                              cognitive arousal 5

. save sem_2fmm, replace
file sem_2fmm.dta saved
```

Notes:

1. You can label the variables and the data, and you can add notes just as you would to any dataset.
2. You save and use SSD just as you save and use any dataset.

Listing the SSD

```
. ssd list
Observations = 216
Means undefined; assumed to be 0
Variances implicitly defined; they are the diagonal of the covariance
matrix.
Covariances:
       a1         a2         a3         a4         a5         c1         c2
2038.035
1631.766   1932.163
1393.567   1336.871   1313.809
1657.299   1647.164   1273.261   2034.216
 1721.83   1688.292   1498.401   1677.767   2061.875
 659.795    702.969    585.019    656.527    775.118    630.518
 779.519    790.448    653.734    764.755    871.211    500.128    741.767
 899.912    879.179    750.037    897.441   1008.287    648.935     679.97
 775.235    739.157    659.867     751.86    895.616    550.476     603.95
  821.17    785.419    669.951    802.825    877.681    491.042    574.775

       c3         c4         c5
1087.409
 677.865    855.272
 686.391     622.83    728.674
```

Also see

[SEM] **example 3** — Two-factor measurement model

[SEM] **ssd** — Making summary statistics data

Title

example 3 — Two-factor measurement model

Description

The multiple-factor measurement model is demonstrated using summary statistics dataset `sem_2fmm.dta`:

```
. use http://www.stata-press.com/data/r12/sem_2fmm
(Affective and cognitive arousal)

. ssd describe

  Summary statistics data from
  http://www.stata-press.com/data/r12/sem_2fmm.dta
    obs:          216                 Affective and cognitive arousal
   vars:           10                 25 May 2011 10:11
                                      (_dta has notes)

  variable name                 variable label

  a1                            affective arousal 1
  a2                            affective arousal 2
  a3                            affective arousal 3
  a4                            affective arousal 4
  a5                            affective arousal 5
  c1                            cognitive arousal 1
  c2                            cognitive arousal 2
  c3                            cognitive arousal 3
  c4                            cognitive arousal 4
  c5                            cognitive arousal 5

. notes
_dta:
  1.  Summary statistics data containing published covariances from Thomas O.
      Williams, Ronald C. Eaves, and Cynthia Cox, 2 Apr 2002, "Confirmatory
      factor analysis of an instrument designed to measure affective and
      cognitive arousal", _Educational and Psychological Measurement_, vol. 62
      no. 2, 264-283.
  2.  a1-a5 report scores from 5 miniscales designed to measure affective
      arousal.
  3.  c1-c5 report scores from 5 miniscales designed to measure cognitive
      arousal.
  4.  The series of tests, known as the VST II (Visual Similes Test II) were
      administered to 216 children ages 10 to 12.  The miniscales are sums of
      scores of 5 to 6 items in VST II.
```

See [SEM] **example 2** to learn how we created this summary statistics dataset.

Remarks

Remarks are presented under the following headings:

Fitting multiple-factor measurement models
Displaying standardized results
Obtaining equation-level goodness of fit using estat eqgof

See *Multiple-factor measurement models* in [SEM] **intro 4** for background.

Fitting multiple-factor measurement models

Below we fit the model shown by Kline (2005, 70–74, 184), namely,

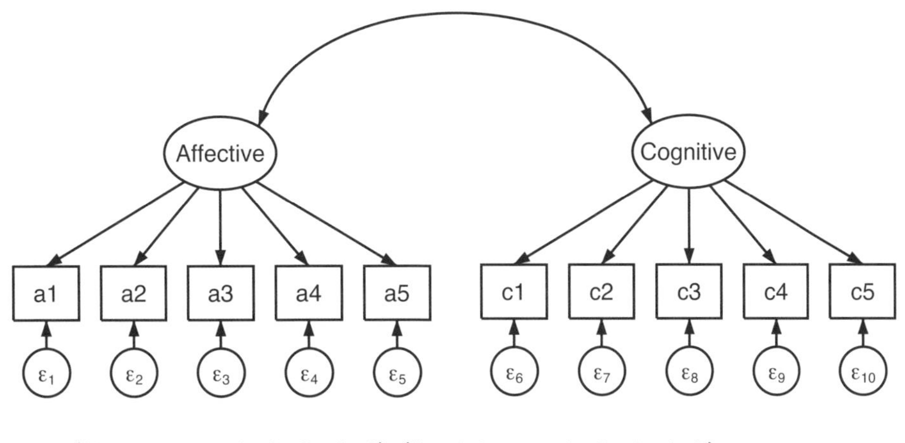


```
. sem (Affective -> a1 a2 a3 a4 a5) (Cognitive -> c1 c2 c3 c4 c5)
Endogenous variables
Measurement:  a1 a2 a3 a4 a5 c1 c2 c3 c4 c5
Exogenous variables
Latent:       Affective Cognitive
Fitting target model:
Iteration 0:   log likelihood = -9542.8803
Iteration 1:   log likelihood = -9539.5505
Iteration 2:   log likelihood = -9539.3856
Iteration 3:   log likelihood = -9539.3851
```

```
Structural equation model                       Number of obs      =       216
Estimation method  = ml
Log likelihood     = -9539.3851

 ( 1)  [a1]Affective = 1
 ( 2)  [c1]Cognitive = 1

------------------------------------------------------------------------------
             |                 OIM
             |      Coef.   Std. Err.      z    P>|z|     [95% Conf. Interval]
-------------+----------------------------------------------------------------
Measurement  |
  a1 <-      |
   Affective |          1  (constrained)
  -----------+----------------------------------------------------------------
  a2 <-      |
   Affective |   .9758098   .0460752    21.18   0.000      .885504    1.066116
  -----------+----------------------------------------------------------------
  a3 <-      |
   Affective |   .8372599   .0355086    23.58   0.000     .7676643    .9068556
  -----------+----------------------------------------------------------------
  a4 <-      |
   Affective |   .9640461   .0499203    19.31   0.000      .866204    1.061888
  -----------+----------------------------------------------------------------
  a5 <-      |
   Affective |   1.063701   .0435751    24.41   0.000     .9782951    1.149107
  -----------+----------------------------------------------------------------
  c1 <-      |
   Cognitive |          1  (constrained)
  -----------+----------------------------------------------------------------
  c2 <-      |
   Cognitive |   1.114702   .0655687    17.00   0.000     .9861901    1.243215
  -----------+----------------------------------------------------------------
  c3 <-      |
   Cognitive |   1.329882   .0791968    16.79   0.000     1.174659    1.485105
  -----------+----------------------------------------------------------------
  c4 <-      |
   Cognitive |   1.172792   .0711692    16.48   0.000     1.033303    1.312281
  -----------+----------------------------------------------------------------
  c5 <-      |
   Cognitive |   1.126356   .0644475    17.48   0.000     1.000041    1.252671
-------------+----------------------------------------------------------------
Variance     |
        e.a1 |   384.1359   43.79119                      307.2194    480.3095
        e.a2 |   357.3524   41.00499                      285.3805    447.4755
        e.a3 |   154.9507   20.09026                      120.1795    199.7822
        e.a4 |   496.4594   54.16323                      400.8838    614.8214
        e.a5 |   191.6857   28.07212                      143.8574    255.4154
        e.c1 |   171.6638   19.82327                       136.894    215.2649
        e.c2 |   171.8055   20.53479                      135.9247    217.1579
        e.c3 |   276.0144   32.33535                      219.3879    347.2569
        e.c4 |   224.1994   25.93412                      178.7197    281.2527
        e.c5 |   146.8655    18.5756                      114.6198    188.1829
   Affective |   1644.463   193.1032                      1306.383    2070.034
   Cognitive |   455.9349   59.11245                      353.6255    587.8439
-------------+----------------------------------------------------------------
Covariance   |
  Affective  |
   Cognitive |   702.0736   85.72272     8.19   0.000     534.0601     870.087
------------------------------------------------------------------------------
LR test of model vs. saturated: chi2(34)  =     88.88, Prob > chi2 = 0.0000
```

Notes:

1. In [SEM] **example 1**, we ran `sem` on raw data. In this example, we run `sem` on summary statistics data. There are no special `sem` options that we need to specify because of this.
2. The estimated coefficients reported above are unstandardized coefficients or, if you prefer, factor loadings.
3. The coefficients listed at the bottom of the coefficient table that start with `e.` are the estimated error variances. They represent the variance of the indicated measurement that is not measured by the respective latent variables.
4. The above results do not match exactly (Kline 2005, 184). If we specified `sem` option `nm1`, results are more likely to match to 3 or 4 digits. The `nm1` option says to divide by $N - 1$ rather than by N in producing variances and covariances.

Displaying standardized results

The output will be easier to interpret if we display standardized values for paths rather than path coefficients. A standardized value is in standard-deviation units. It is the change in one variable given a change in another, both measured in standard-deviation units. We can obtain standardized values by specifying `sem`'s `standardized` option, which we can do when we fit the model or when we replay results:

```
. sem, standardized

Structural equation model                       Number of obs      =        216
Estimation method  = ml
Log likelihood     = -9539.3851

 ( 1)  [a1]Affective = 1
 ( 2)  [c1]Cognitive = 1
```

Standardized	Coef.	OIM Std. Err.	z	P>|z|	[95% Conf.	Interval]
Measurement						
a1 <- Affective	.9003553	.0143988	62.53	0.000	.8721342	.9285765
a2 <- Affective	.9023249	.0141867	63.60	0.000	.8745195	.9301304
a3 <- Affective	.9388883	.0097501	96.29	0.000	.9197784	.9579983
a4 <- Affective	.8687982	.0181922	47.76	0.000	.8331421	.9044543
a5 <- Affective	.9521559	.0083489	114.05	0.000	.9357923	.9685195
c1 <- Cognitive	.8523351	.0212439	40.12	0.000	.8106978	.8939725
c2 <- Cognitive	.8759601	.0184216	47.55	0.000	.8398544	.9120658
c3 <- Cognitive	.863129	.0199624	43.24	0.000	.8240033	.9022547
c4 <- Cognitive	.8582786	.0204477	41.97	0.000	.8182018	.8983554
c5 <- Cognitive	.8930346	.0166261	53.71	0.000	.8604479	.9256212
Variance						
e.a1	.1893602	.0259281			.1447899	.2476506
e.a2	.1858097	.0256021			.1418353	.2434179
e.a3	.1184887	.0183086			.0875289	.1603993
e.a4	.2451896	.0316107			.1904417	.3156764
e.a5	.0933991	.015899			.0669031	.1303885
e.c1	.2735248	.0362139			.2110086	.354563
e.c2	.2326939	.0322732			.1773081	.3053806
e.c3	.2550083	.0344603			.1956717	.3323385
e.c4	.2633578	.0350997			.2028151	.3419733
e.c5	.2024893	.0296954			.1519049	.2699183
Affective	1	.			.	.
Cognitive	1	.			.	.
Covariance						
Affective Cognitive	.8108102	.0268853	30.16	0.000	.758116	.8635045

```
LR test of model vs. saturated: chi2(34)  =     88.88, Prob > chi2 = 0.0000
```

Notes:

1. In addition to obtaining standardized coefficients, the `standardized` option reports estimated error variances as the fraction of the variance that is unexplained. Error variances were previously unintelligible numbers such as 384.1359 and 357.3524. Now they are 0.189 and 0.186.

2. Also listed in the `sem` output are variances of latent variables. In the previous output, latent variable `Affective` had variance 1,644.46 with standard error 193. In the standardized output, it has variance 1 with standard error missing. The variances of the latent variables are standardized to 1, and obviously, being a normalization, there is no corresponding standard error.

3. We can now see at the bottom of the coefficient table that affective and cognitive arousal are correlated 0.81 because standardized covariances are correlation coefficients.

4. The standardized coefficients for this model can be interpreted as the correlation coefficients between the indicator and the latent variable because each indicator measures only one factor. For instance, the standardized path coefficient `a1<-Affective` is 0.90, meaning the correlation between `a1` and `Affective` is 0.90.

Obtaining equation-level goodness of fit using estat eqgof

That the correlation between `a1` and `Affective` is 0.90 implies that the fraction of the variance of `a1` explained by `Affective` is $0.90^2 = 0.81$, and left unexplained is $1 - 0.81 = 0.19$. Instead of manually calculating the proportion of variance explained by indicators, we can use the `estat eqgof` command:

```
. estat eqgof
Equation-level goodness of fit
```

depvars	fitted	Variance predicted	residual	R-squared	mc	mc2
observed						
a1	2028.598	1644.463	384.1359	.8106398	.9003553	.8106398
a2	1923.217	1565.865	357.3524	.8141903	.9023249	.8141903
a3	1307.726	1152.775	154.9507	.8815113	.9388883	.8815113
a4	2024.798	1528.339	496.4594	.7548104	.8687982	.7548104
a5	2052.328	1860.643	191.6857	.9066009	.9521559	.9066009
c1	627.5987	455.9349	171.6638	.7264752	.8523351	.7264752
c2	738.3325	566.527	171.8055	.7673061	.8759601	.7673061
c3	1082.374	806.3598	276.0144	.7449917	.863129	.7449917
c4	851.311	627.1116	224.1994	.7366422	.8582786	.7366422
c5	725.3002	578.4346	146.8655	.7975107	.8930346	.7975107
overall				.9949997		

```
mc  = correlation between depvar and its prediction
mc2 = mc^2 is the Bentler-Raykov squared multiple correlation coefficient
```

Notes:

1. `fitted` reports the fitted variance of each of the endogenous variables whether observed or latent. In this case, we have observed endogenous variables.

2. `predicted` reports the variance of the predicted value of each endogenous variable.

3. `residual` reports the leftover residual variance.

4. `R-squared` reports R^2, the fraction of variance explained by each indicator. The fraction of the variance of `Affective` explained by `a1` is 0.81, just as we calculated by hand above, at the beginning of this section. The overall R^2 is also called the coefficient of determination.

5. `mc` stands for multiple correlation and `mc2` stands for multiple-correlation squared. `R-squared`, `mc`, and `mc2` all report the relatedness of the indicated dependent variable with the model's linear prediction. In recursive models, all three statistics are really the same number. `mc` is equal to the square root of `R-squared`, and `mc2` is equal to `R-squared`.

   In nonrecursive models, these three statistics are different and each can have problems. `R-squared` and `mc` can actually become negative! That does not mean the model has negative predictive power or that it might not even have reasonable predictive power. $\texttt{mc2} = \texttt{mc}^2$ is recommended by Bentler and Raykov (2000) to be used instead of `R-squared` for nonrecursive systems.

In [SEM] **example 4**, we examine the goodness-of-fit statistics for this model.

In [SEM] **example 5**, we examine modification indices for this model.

Also see

[SEM] **example 1** — Single-factor measurement model

[SEM] **example 2** — Creating a dataset from published covariances

[SEM] **example 20** — Two-factor measurement model by group

[SEM] **example 26** — Fitting a model using data missing at random

[SEM] **sem** — Structural equation model estimation command

[SEM] **estat eqgof** — Equation-level goodness-of-fit statistics

Title

example 4 — Goodness-of-fit statistics

Description

We demonstrate `estat gof`. See [SEM] **intro 6** and see [SEM] **estat gof**.

This example picks up where [SEM] **example 3** left off:

```
. use http://www.stata-press.com/data/r12/sem_2fmm
. sem (Affective -> a1 a2 a3 a4 a5) (Cognitive -> c1 c2 c3 c4 c5)
```

Remarks

When we fit this model in [SEM] **example 3**, at the bottom of the output, we saw

```
. sem (Affective -> a1 a2 a3 a4 a5) (Cognitive -> c1 c2 c3 c4 c5)
  (output omitted)
LR test of model vs. saturated: chi2(34)   =     88.88, Prob > chi2 = 0.0000
```

Most texts refer to this test against the saturated model as the "model χ^2 test".

These results indicate poor goodness of fit; see [SEM] **example 1**. The default goodness-of-fit statistic reported by `sem`, however, can be overly influenced by sample size, correlations, variance unrelated to the model, and multivariate nonnormality (Kline 2011, 201).

Goodness of fit in cases of `sem` is a measure of how well you fit the observed moments, which in this case are the covariances between all pairs of `a1`, ..., `a5`, `c1`, ..., `c5`. In a measurement model, the assumed underlying causes are unobserved, and in this example, those unobserved causes are the latent variables `Affective` and `Cognitive`. It may be reasonable to assume that the observed `a1`, ..., `a5`, `c1`, ..., `c5` can be filtered through imagined variables `Affective` and `Cognitive`, but that can be reasonable only if not too much information contained in the original variables is lost. Thus goodness-of-fit statistics are of great interest to those fitting measurement models. Goodness-of-fit statistics are of far less interest when all variables in the model are observed.

Other goodness-of-fit statistics are available.

```
. estat gof, stats(all)
```

Fit statistic	Value	Description
Likelihood ratio		
chi2_ms(34)	88.879	model vs. saturated
p > chi2	0.000	
chi2_bs(45)	2467.161	baseline vs. saturated
p > chi2	0.000	
Population error		
RMSEA	0.080	Root mean squared error of approximation
90% CI, lower bound	0.065	
upper bound	0.109	
pclose	0.004	Probability RMSEA <= 0.05
Information criteria		
AIC	19120.770	Akaike's information criterion
BIC	19191.651	Bayesian information criterion
Baseline comparison		
CFI	0.977	Comparative fit index
TLI	0.970	Tucker-Lewis index
Size of residuals		
SRMR	0.022	Standardized root mean squared residual
CD	0.995	Coefficient of determination

Notes:

1. Desirable values vary from test to test.

2. We asked for all the goodness-of-fit tests. We could have obtained specific tests from the above output by specifying the appropriate option; see [SEM] **estat gof**.

3. Under likelihood ratio, `estat gof` reports two tests. The first is a repeat of the model χ^2 test reported at the bottom of the `sem` output. The saturated model is the model that fits the covariances perfectly. We can reject at the 5% level (or any other level) that the model fits as well as the saturated model.

   The second test is a baseline versus saturated comparison. The baseline model includes the mean and variances of all observed variables plus the covariances of all observed exogenous variables. Different authors define the baseline differently. We can reject at the 5% level (or any other level) that the baseline model fits as well as the saturated model.

4. Under population error, the RMSEA value is reported along with the lower and upper bounds of its 90% confidence interval. Most interpreters of this test check whether the lower bound is below 0.05 or the upper bound is above 0.10. If the lower bound is below 0.05, then they would not reject the hypothesis that the fit is close. If the upper bound is above 0.10, they would not reject the hypothesis that the fit is poor. The logic is to perform one test on each end of the 90% confidence interval and thus have 95% confidence in the result. This model's fit is not close, and its upper limit is just over the bounds of being considered poor.

   Pclose, a commonly used word in reference to this test, is the probability that the RMSEA value is less than 0.05, interpreted as the probability that the predicted moments are close to the moments in the population. This model's fit is not close.

5. Under information criteria are reported AIC and BIC, which contain little information by themselves but are often used to compare models. Smaller values are considered better.
6. Under baseline comparison are reported CFI and TLI, two indices such that a value close to 1 indicates a good fit. TLI is also known as the nonnormed fit index.
7. Under size of residuals is reported the standardized root mean squared residual (SRMR) and the coefficient of determination (CD).

   A perfect fit corresponds to an SRMR of 0, and a good fit corresponds to a "small" value, considered by some to be limited at 0.08. The model fits well by this standard.

   The CD is like an R^2 for the whole model. A value close to 1 indicates a good fit.

`estat gof` provides multiple goodness-of-fit statistics because, across fields, different researchers use different statistics. You should not print them all and look for the one reporting the result you seek.

Also see

[SEM] **example 3** — Two-factor measurement model

[SEM] **example 21** — Group-level goodness of fit

[SEM] **estat gof** — Goodness-of-fit statistics

Title

example 5 — Modification indices

Description

We demonstrate the use of `estat mindices`; see [SEM] **intro 6** and see [SEM] **estat mindices**.

This example picks up where [SEM] **example 3** left off:

```
. use http://www.stata-press.com/data/r12/sem_2fmm
. sem (Affective -> a1 a2 a3 a4 a5) (Cognitive -> c1 c2 c3 c4 c5)
```

Remarks

When we fit this model in [SEM] **example 4**, we allowed the latent variables to be correlated. We typed

```
. sem (Affective -> a1 a2 a3 a4 a5) (Cognitive -> c1 c2 c3 c4 c5)
```

and by default in the command language, latent exogenous variables are assumed to be correlated unless we specify otherwise. Had we used the GUI, the latent exogenous variables would have been assumed to be uncorrelated unless we had drawn the curved path between them.

The original authors who collected these data analyzed them assuming no covariance, which we could obtain by typing

```
. sem (Affective -> a1 a2 a3 a4 a5) (Cognitive -> c1 c2 c3 c4 c5), ///
                                          cov(Affective*Cognitive@0)
```

It was Kline (2005, 70–74, 184) who allowed the covariance. Possibly he did that after looking at the modification indices.

The modification indices report statistics on all omitted paths. Let's begin with the model without the covariance:

```
. sem (Affective -> a1 a2 a3 a4 a5) (Cognitive -> c1 c2 c3 c4 c5),
>                                    cov(Affective*Cognitive@0)
  (output omitted )
. estat mindices
Modification indices
```

	MI	df	P>MI	EPC	Standard EPC
Measurement					
a5 <-					
Cognitive	8.059	1	0.00	.1604476	.075774
c5 <-					
Affective	5.885	1	0.02	.0580897	.087733
Covariance					
e.a1					
e.a4	5.767	1	0.02	84.81133	.1972802
e.a5	7.597	1	0.01	-81.82092	-.2938627
e.a2					
e.a4	14.300	1	0.00	129.761	.3110565
e.c4	4.071	1	0.04	-45.44807	-.1641344
e.a3					
e.a4	21.183	1	0.00	-116.8181	-.4267012
e.a5	25.232	1	0.00	118.4674	.6681337
e.a5					
e.c4	4.209	1	0.04	39.07999	.184049
e.c1					
e.c3	11.326	1	0.00	66.3965	.3098331
e.c5	8.984	1	0.00	-47.31483	-.2931597
e.c3					
e.c4	12.668	1	0.00	-80.98353	-.333871
e.c4					
e.c5	4.483	1	0.03	38.6556	.2116015
Affective					
Cognitive	128.482	1	0.00	704.4469	.8094959

```
EPC = expected parameter change
```

Notes:

1. Four columns of results are reported.

   a. MI stands for modification index and is an approximation to the change in the model's goodness-of-fit χ^2 if the path were added.

   b. df stands for degrees of freedom and is the number that would be added to d of the $\chi^2(d)$.

   c. P>MI is the value of the significance of $\chi^2(\texttt{df})$.

   d. EPC stands for expected parameter change and is an approximation to the value of the parameter if it were not constrained to 0. It is reported in unstandardized (column 3) and standardized (column 4) units.

2. There are lots of significant omitted paths in the above output.

3. Paths are listed only if the modification index is significant at the 0.05 level, corresponding to $\chi^2(1)$ value 3.8414588. You may specify the `minchi2()` option to use different $\chi^2(1)$ values. Specify `minchi2(0)` if you wish to see all tests.
4. The omitted path between `Affective` and `Cognitive` has the largest change in χ^2 observed. Perhaps this is why Kline (2005, 70–74, 184) allowed a covariance between the two latent variables. The standardized `EPC` reports the relaxed-constraint correlation value, which is the value reported for the unconstrained correlation path in [SEM] **example 3**.

Another way of dealing with this significant result would be to add a direct path between the variables, but that perhaps would have invalidated the theory being proposed. The original authors instead proposed a second order model postulating that `Affective` and `Cognitive` are themselves measurements of another latent variable that might be called `Arousal`.

Also see

[SEM] **example 3** — Two-factor measurement model

[SEM] **estat mindices** — Modification indices

Title

example 6 — Linear regression

Description

Linear regression is demonstrated using `auto.dta`:

```
. sysuse auto, clear
(1978 Automobile Data)
```

See *Structural models 1: Linear regression* in [SEM] **intro 4** for background.

Remarks

The first two examples in [R] **regress** are

```
. regress mpg weight c.weight#c.weight foreign
. regress, beta
```

This model corresponds to

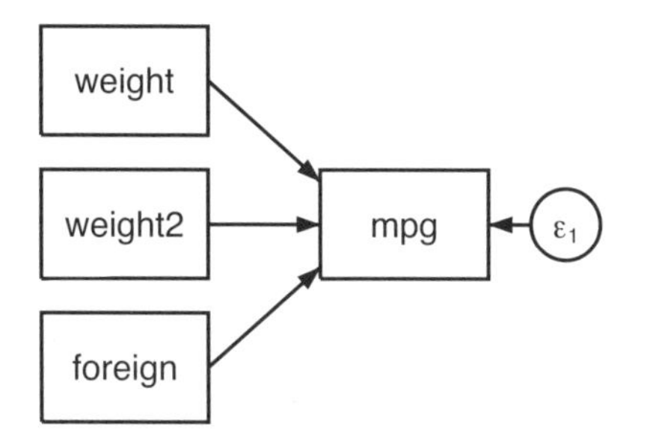

To fit this model using sem, we type

```
. generate weight2 = weight^2
. sem (mpg <- weight weight2 foreign)
Endogenous variables
Observed:  mpg
Exogenous variables
Observed:  weight weight2 foreign
Fitting target model:
Iteration 0:   log likelihood = -1909.8206
Iteration 1:   log likelihood = -1909.8206
Structural equation model                       Number of obs      =        74
Estimation method  = ml
Log likelihood     = -1909.8206
```

	Coef.	OIM Std. Err.	z	P>|z|	[95% Conf.	Interval]
Structural						
mpg <-						
weight	-.0165729	.0038604	-4.29	0.000	-.0241392	-.0090067
weight2	1.59e-06	6.08e-07	2.62	0.009	4.00e-07	2.78e-06
foreign	-2.2035	1.03022	-2.14	0.032	-4.222695	-.1843056
_cons	56.53884	6.027559	9.38	0.000	44.72504	68.35264
Variance						
e.mpg	10.19332	1.675772			7.385485	14.06865

```
LR test of model vs. saturated: chi2(1)   =      0.00, Prob > chi2 = 1.0000
```

Notes:

1. We wished to include variable weight^2 in our model. Because sem does not allow Stata's factor-variable notation, we first had to generate new variable weight2.
2. Reported coefficients match those reported by regress.
3. Reported standard errors (SEs) differ slightly from those reported by regress. For instance, the SE for foreign is reported here as 1.03, whereas regress reported 1.06. SEM is an asymptotic estimator, and sem divides variances and covariances by $N = 74$, the number of observations. regress provides unbiased finite-sample estimates and divides by $N - k - 1 = 74 - 3 - 1 = 70$. Note that $1.03\sqrt{74/70} = 1.06$.
4. sem reports z statistics whereas regress reports t statistics.
5. Reported confidence intervals differ slightly between sem and regress because of the $(N - k - 1)/N$ issue.
6. sem reports the point estimate of e.mpg as 10.19332; regress reports the root MSE as 3.2827, and $\sqrt{10.19332 \times 74/70} = 3.2827$.

To obtain standardized coefficients from regress, you specify the beta option. To obtain standardized coefficients from sem, you specify the standardized option.

```
. sem, standardized

Structural equation model                       Number of obs      =        74
Estimation method  = ml
Log likelihood     = -1909.8206

------------------------------------------------------------------------------
             |                 OIM
Standardized |      Coef.   Std. Err.      z    P>|z|     [95% Conf. Interval]
-------------+----------------------------------------------------------------
Structural   |
  mpg <-     |
      weight |  -2.226321   .4950378    -4.50   0.000    -3.196577   -1.256064
     weight2 |    1.32654    .498261     2.66   0.008     .3499662    2.303113
     foreign |    -.17527   .0810378    -2.16   0.031    -.3341011   -.0164389
       _cons |   9.839209   .9686872    10.16   0.000     7.940617     11.7378
-------------+----------------------------------------------------------------
Variance     |
       e.mpg |    .308704   .0482719                      .2272168    .4194152
------------------------------------------------------------------------------
LR test of model vs. saturated: chi2(1)   =      0.00, Prob > chi2 = 1.0000
```

`regress` simply reports standardized coefficients in an extra column. All other results are reported in unstandardized form. `sem` updates the entire output with the standardized values.

Also see

[SEM] **example 12** — Seemingly unrelated regression

[SEM] **sem** — Structural equation model estimation command

Title

example 7 — Nonrecursive structural model

Description

To demonstrate a nonrecursive structural model with all variables observed, we use data from Duncan, Haller, and Portes (1968):

```
. use http://www.stata-press.com/data/r12/sem_sm1
(Structural model with all observed values)

. ssd describe

  Summary statistics data from
  http://www.stata-press.com/data/r12/sem_sm1.dta
    obs:            329                Structural model with all obse..
   vars:             10                25 May 2011 10:13
                                       (_dta has notes)
  ------------------------------------------------------------------------
  variable name                 variable label
  ------------------------------------------------------------------------
  r_intel                       respondent's intelligence
  r_parasp                      respondent's parental aspiration
  r_ses                         respondent's family socioeconomic status
  r_occasp                      respondent's occupational aspiration
  r_educasp                     respondent's educational aspiration
  f_intel                       friend's intelligence
  f_parasp                      friend's parental aspiration
  f_ses                         friend's family socioeconomic status
  f_occasp                      friend's occupational aspiration
  f_educasp                     friend's educational aspiration
  ------------------------------------------------------------------------

. notes

_dta:
  1.  Summary statistics data from Duncan, O.D., Haller, A.O., and Portes, A.,
      1968, "Peer Influences on Aspirations:  A Reinterpretation", _American
      Journal of Sociology_ 74, 119-137.
  2.  The data contain 329 boys with information on five variables and the same
      information for each boy's best friend.
```

If you typed `ssd status`, you would learn that this dataset contains the correlation matrix only. Variances (standard deviations) and means are undefined. Thus we need to use this dataset cautiously. It is always better if you enter the variances and means if you have them.

That these data are the correlations only will not matter for how we will use them.

Remarks

See *Structural models 2: Dependencies between endogenous variables* in [SEM] **intro 4** for background.

Remarks are presented under the following headings:

Fitting the model
Checking stability using estat stable
Reporting total, direct, and indirect effects using estat teffects

Fitting the model

In the referenced paper above, the authors fit the following model:

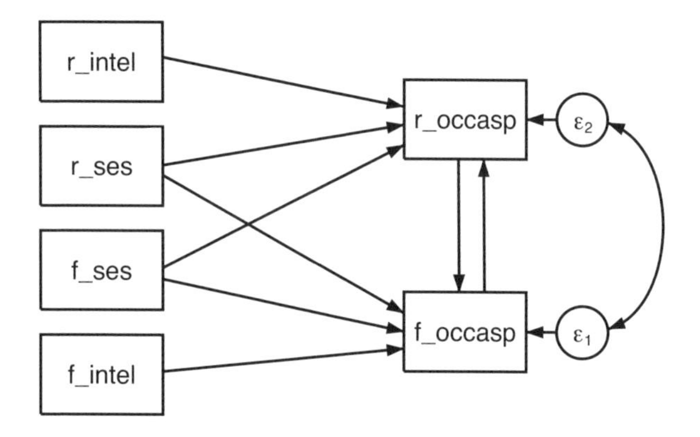

```
. sem (r_occasp <- f_occasp r_intel r_ses f_ses)
>     (f_occasp <- r_occasp f_intel f_ses r_ses),
>                          cov(e.r_occasp*e.f_occasp) standardized
Endogenous variables

Observed:  r_occasp f_occasp

Exogenous variables

Observed:  r_intel r_ses f_ses f_intel

Fitting target model:

Iteration 0:   log likelihood = -2617.0489
Iteration 1:   log likelihood = -2617.0489

Structural equation model                       Number of obs      =       329
Estimation method  = ml
Log likelihood     = -2617.0489

------------------------------------------------------------------------------
             |                 OIM
Standardized |      Coef.   Std. Err.      z    P>|z|     [95% Conf. Interval]
-------------+----------------------------------------------------------------
Structural   |
  r_occ~p <- |
    f_occasp |   .2773441   .1281904     2.16   0.031     .0260956    .5285926
     r_intel |   .2854766        .05     5.71   0.000     .1874783    .3834748
       r_ses |   .1570082   .0520841     3.01   0.003     .0549252    .2590912
       f_ses |   .0973327    .060153     1.62   0.106     -.020565    .2152304
  -----------+----------------------------------------------------------------
  f_occ~p <- |
    r_occasp |   .2118102    .156297     1.36   0.175    -.0945264    .5181467
       r_ses |   .0794194   .0587732     1.35   0.177    -.0357739    .1946127
       f_ses |   .1681772   .0537199     3.13   0.002      .062888    .2734663
     f_intel |   .3693682   .0525924     7.02   0.000     .2662891    .4724474
-------------+----------------------------------------------------------------
Variance     |
  e.r_occasp |   .6889244   .0399973                      .6148268    .7719519
  e.f_occasp |   .6378539    .039965                      .5641425    .7211964
-------------+----------------------------------------------------------------
Covariance   |
  e.r_occasp |
  e.f_occasp |  -.2325666   .2180087    -1.07   0.286    -.6598558    .1947227
------------------------------------------------------------------------------
LR test of model vs. saturated: chi2(0)   =      0.00, Prob > chi2 =      .
```

Notes:

1. We specified the `standardized` option, but in this case that did not matter much because these data are based on the correlation coefficients only, so standardized values are equal to unstandardized values. The exception is the correlation between the latent endogenous variables, as reflected in the correlation of their errors, and we wanted to show that results match those in the original paper.

2. Nearly all results match those in the original paper. The authors normalized the errors to have a variance of 1; `sem` normalizes the paths from the errors to have coefficient 1. While you can apply most normalizing constraints any way you wish, `sem` restricts errors to have path coefficients of 1 and this cannot be modified. You could, however, prove to yourself that `sem` would produce the same variances as the authors produced by typing

    ```
    . sem, coeflegend
    . display sqrt(_b[var(e.r_occasp):_cons]
    . display sqrt(_b[var(e.f_occasp):_cons]
    ```

because the coefficients would be the standard deviations of the errors estimated without the variance-1 constraint. Thus all results match. We replayed results by using the `coeflegend` option so that we would know what to type to refer to the two error variances, namely, `_b[var(e.r_occasp):_cons]` and `_b[var(e.f_occasp):_cons]`.

Checking stability using estat stable

```
. estat stable
Stability analysis of simultaneous equation systems
   Eigenvalue stability condition
```

Eigenvalue	Modulus
-.2423722	.242372
.2423722	.242372

```
   stability index =  .2423722
   All the eigenvalues lie inside the unit circle.
   SEM satisfies stability condition.
```

Notes:

1. `estat stable` is for use on nonrecursive models. Recursive models are by design stable.
2. Stability concerns whether the parameters of the model are such that the model would blow up if it were operated over and over again. If the results are found not to be stable, then that casts questions about the validity of the model.
3. The stability is the maximum of the moduli, and the moduli are the absolute values of the eigenvalues. Usually the two eigenvalues are not identical, but it is a property of this model that they are.
4. If the stability index is less than one, then the reported estimates yield a stable model.

In the next section, we use `estat teffects` to estimate total effects. That is appropriate only if the model is stable, as we find that it is.

Reporting total, direct, and indirect effects using estat teffects

```
. estat teffects
```

Direct effects

	Coef.	OIM Std. Err.	z	P>|z|	[95% Conf.	Interval]
Structural						
r_occ~p <-						
r_occasp	0	(no path)				
f_occasp	.2773441	.1287622	2.15	0.031	.0249748	.5297134
r_intel	.2854766	.0522001	5.47	0.000	.1831662	.3877869
r_ses	.1570082	.052700	2.98	0.003	.0536534	.260363
f_ses	.0973327	.0603699	1.61	0.107	-.0200001	.2156555
f_intel	0	(no path)				
f_occ~p <-						
r_occasp	.2118102	.1563958	1.35	0.176	-.09472	.5183404
f_occasp	0	(no path)				
r_intel	0	(no path)				
r_ses	.0794194	.0589095	1.35	0.178	-.0360411	.1948799
f_ses	.1681772	.0543854	3.09	0.002	.0615838	.2747705
f_intel	.3693682	.0557939	6.62	0.000	.2600142	.4787223

Indirect effects

	Coef.	OIM Std. Err.	z	P>|z|	[95% Conf.	Interval]
Structural						
r_occ~p <-						
r_occasp	.0624106	.0460825	1.35	0.176	-.0279096	.1527307
f_occasp	.0173092	.0080361	2.15	0.031	.0015587	.0330597
r_intel	.0178168	.0159383	1.12	0.264	-.0134217	.0490552
r_ses	.0332001	.0204531	1.62	0.105	-.0068872	.0732875
f_ses	.0556285	.0292043	1.90	0.057	-.0016109	.112868
f_intel	.1088356	.052243	2.08	0.037	.0064411	.21123
f_occ~p <-						
r_occasp	.0132192	.0097608	1.35	0.176	-.0059115	.0323499
f_occasp	.0624106	.0289753	2.15	0.031	.0056201	.1192011
r_intel	.0642406	.0490164	1.31	0.190	-.0318298	.160311
r_ses	.0402881	.0315496	1.28	0.202	-.021548	.1021242
f_ses	.0323987	.0262124	1.24	0.216	-.0189765	.083774
f_intel	.0230525	.0202112	1.14	0.254	-.0165607	.0626657

```
Total effects

------------------------------------------------------------------------------
             |                 OIM
             |      Coef.   Std. Err.      z    P>|z|     [95% Conf. Interval]
-------------+----------------------------------------------------------------
Structural   |
  r_occ~p <- |
    r_occasp |   .0624106   .0460825     1.35   0.176    -.0279096    .1527307
    f_occasp |   .2946533   .1367983     2.15   0.031     .0265335    .5627731
     r_intel |   .3032933   .0509684     5.95   0.000     .2033971    .4031896
       r_ses |   .1902083    .050319     3.78   0.000      .091585    .2888317
       f_ses |   .1529612    .050844     3.01   0.003     .0533089    .2526136
     f_intel |   .1088356    .052243     2.08   0.037     .0064411      .21123
-------------+----------------------------------------------------------------
  f_occ~p <- |
    r_occasp |   .2250294   .1661566     1.35   0.176    -.1006315    .5506903
    f_occasp |   .0624106   .0289753     2.15   0.031     .0056201    .1192011
     r_intel |   .0642406   .0490164     1.31   0.190    -.0318298     .160311
       r_ses |   .1197074   .0483919     2.47   0.013     .0248611    .2145537
       f_ses |   .2005759   .0488967     4.10   0.000       .10474    .2964118
     f_intel |   .3924207   .0502422     7.81   0.000     .2939478    .4908936
------------------------------------------------------------------------------
```

Notes:

1. In the path diagram we drew for this model, you can see that the intelligence of the respondent `r_intel` has both direct and indirect effects on the occupational aspiration of the respondent `r_occasp`. The tables above reveal that

$$0.303 = 0.285 + 0.018$$

where 0.285 is the direct effect and 0.018 is the indirect effect.

Also see

[SEM] **example 8** — Testing that coefficients are equal, and constraining them

[SEM] **sem** — Structural equation model estimation command

[SEM] **estat stable** — Check stability of nonrecursive system

[SEM] **estat teffects** — Decomposition of effects into total, direct, and indirect

Title

example 8 — Testing that coefficients are equal, and constraining them

Description

This example continues from where [SEM] **example 7** left off, where we typed

```
. use http://www.stata-press.com/data/r12/sem_sm1

. ssd describe

  notes

. sem (r_occasp <- f_occasp r_intel r_ses f_ses)  ///
      (f_occasp <- r_occasp f_intel f_ses r_ses), ///
                   cov(e.r_occasp*e.f_occasp) standardized

. estat stable

. estat teffects
```

Remarks

Remarks are presented under the following headings:

Using test to evaluate adding constraints
Refitting the model with added constraints
Using estat scoretests to test whether constraints can be relaxed

That is, we want to show you (1) how to evaluate potential constraints after estimation, (2) how to fit a model with constraints, and (3) how to evaluate enforced constraints after estimation.

Obviously, in a real analysis, if you did (1) there would be no reason to do (3), and vice versa.

Using test to evaluate adding constraints

In this model of respondents and corresponding friends, it would be surprising if the coefficients relating friends' characteristics to respondents' occupational aspirations and vice versa were not equal. It would also be surprising if coefficients relating respondents' characteristics to his occupational aspirations were not equal to those of friends' characteristics to his occupational aspirations. The paths that we suspect should be equal are

```
r_intel  -> r_occasp          f_intel  -> f_occasp
r_ses    -> r_occasp          f_ses    -> f_occasp
f_ses    -> r_occasp          r_ses    -> f_occasp
f_occasp -> r_occasp          r_occasp -> f_occasp
```

You are about to learn that to test whether those paths have equal coefficients, you type

```
. test (_b[r_occasp:r_intel ]==_b[f_occasp:f_intel ])  ///
       (_b[r_occasp:r_ses   ]==_b[f_occasp:f_ses   ])  ///
       (_b[r_occasp:f_ses   ]==_b[f_occasp:r_ses   ])  ///
       (_b[r_occasp:f_occasp]==_b[f_occasp:r_occasp])
```

In Stata, _b[] is how one accesses the estimated parameters. It is difficult to remember what the names are. To determine the names of the parameters, replay the sem results with the coeflegend option:

```
. sem, coeflegend

Structural equation model                       Number of obs      =        329
Estimation method  = ml
Log likelihood     = -2617.0489

------------------------------------------------------------------------------
                       |      Coef.  Legend
-----------------------+------------------------------------------------------
Structural             |
  r_occ~p <-           |
       f_occasp        |   .2773441  _b[r_occasp:f_occasp]
        r_intel        |   .2854766  _b[r_occasp:r_intel]
          r_ses        |   .1570082  _b[r_occasp:r_ses]
          f_ses        |   .0973327  _b[r_occasp:f_ses]
-----------------------+------------------------------------------------------
  f_occ~p <-           |
       r_occasp        |   .2118102  _b[f_occasp:r_occasp]
          r_ses        |   .0794194  _b[f_occasp:r_ses]
          f_ses        |   .1681772  _b[f_occasp:f_ses]
        f_intel        |   .3693682  _b[f_occasp:f_intel]
-----------------------+------------------------------------------------------
Variance               |
  e.r_occasp           |   .6868304  _b[var(e.r_occasp):_cons]
  e.f_occasp           |   .6359151  _b[var(e.f_occasp):_cons]
-----------------------+------------------------------------------------------
Covariance             |
  e.r_occasp           |
  e.f_occasp           |  -.1536992  _b[cov(e.r_occasp,e.f_occasp):_cons]
------------------------------------------------------------------------------
LR test of model vs. saturated: chi2(0)   =      0.00, Prob > chi2 =      .
```

With the parameter names at hand, to perform the test, we can type

```
. test (_b[r_occasp:r_intel ]==_b[f_occasp:f_intel ])
>      (_b[r_occasp:r_ses   ]==_b[f_occasp:f_ses   ])
>      (_b[r_occasp:f_ses   ]==_b[f_occasp:r_ses   ])
>      (_b[r_occasp:f_occasp]==_b[f_occasp:r_occasp])

 ( 1)  [r_occasp]r_intel - [f_occasp]f_intel = 0
 ( 2)  [r_occasp]r_ses - [f_occasp]f_ses = 0
 ( 3)  [r_occasp]f_ses - [f_occasp]r_ses = 0
 ( 4)  [r_occasp]f_occasp - [f_occasp]r_occasp = 0

           chi2(  4) =    1.61
         Prob > chi2 =    0.8062
```

We cannot reject the constraint, just as we expected.

Refitting the model with added constraints

We could refit the model with these constraints by typing

```
. sem (r_occasp <- f_occasp@b1  r_intel@b2  r_ses@b3  f_ses@b4)
>     (f_occasp <- r_occasp@b1  f_intel@b2  f_ses@b3  r_ses@b4),
>                               cov(e.r_occasp*e.f_occasp)

Endogenous variables

Observed:  r_occasp f_occasp

Exogenous variables

Observed:  r_intel r_ses f_ses f_intel

Fitting target model:

Iteration 0:   log likelihood = -2617.8735
Iteration 1:   log likelihood = -2617.8705
Iteration 2:   log likelihood = -2617.8705

Structural equation model                       Number of obs      =       329
Estimation method  = ml
Log likelihood     = -2617.8705

 ( 1)  [r_occasp]f_occasp - [f_occasp]r_occasp = 0
 ( 2)  [r_occasp]r_intel - [f_occasp]f_intel = 0
 ( 3)  [r_occasp]r_ses - [f_occasp]f_ses = 0
 ( 4)  [r_occasp]f_ses - [f_occasp]r_ses = 0

------------------------------------------------------------------------------
             |                 OIM
             |      Coef.   Std. Err.      z    P>|z|     [95% Conf. Interval]
-------------+----------------------------------------------------------------
Structural   |
  r_occ~p <- |
    f_occasp |   .2471578   .1024504     2.41   0.016     .0463588    .4479568
     r_intel |   .3271847   .0407973     8.02   0.000     .2472234    .4071459
       r_ses |   .1635056   .0380582     4.30   0.000     .0889129    .2380984
       f_ses |    .088364   .0427106     2.07   0.039     .0046529    .1720752
  -----------+----------------------------------------------------------------
  f_occ~p <- |
    r_occasp |   .2471578   .1024504     2.41   0.016     .0463588    .4479568
       r_ses |    .088364   .0427106     2.07   0.039     .0046529    .1720752
       f_ses |   .1635056   .0380582     4.30   0.000     .0889129    .2380984
     f_intel |   .3271847   .0407973     8.02   0.000     .2472234    .4071459
-------------+----------------------------------------------------------------
Variance     |
  e.r_occasp |   .6884513   .0538641                      .5905757    .8025477
  e.f_occasp |   .6364713   .0496867                      .5461715    .7417005
-------------+----------------------------------------------------------------
Covariance   |
  e.r_occasp |
  e.f_occasp |  -.1582175   .1410111    -1.12   0.262    -.4345942    .1181592
------------------------------------------------------------------------------
LR test of model vs. saturated: chi2(4)   =      1.64, Prob > chi2 = 0.8010
```

Using estat scoretests to test whether constraints can be relaxed

```
. estat scoretests
(no score tests to report; all chi2 values less than 3.841458820694123)
```

No tests were reported because no tests were individually significant at the 5% level. We can obtain all the individual tests by adding the minchi2(0) option, which we can abbreviate to min(0):

```
. estat scoretests, min(0)
Score tests for linear constraints
  ( 1)  [r_occasp]f_occasp - [f_occasp]r_occasp = 0
  ( 2)  [r_occasp]r_intel - [f_occasp]f_intel = 0
  ( 3)  [r_occasp]r_ses - [f_occasp]f_ses = 0
  ( 4)  [r_occasp]f_ses - [f_occasp]r_ses = 0
```

	chi2	df	P>chi2
(1)	0.014	1	0.91
(2)	1.225	1	0.27
(3)	0.055	1	0.81
(4)	0.136	1	0.71

Notes:

1. When we began this example, we used test to evaluate potential constraints that we were considering. We obtained an overall $\chi^2(4)$ statistic of 1.61 and thus could not reject the constraints at any reasonable level.
2. We then refit the model using those constraints.
3. For pedantic reasons, now we use estat scoretests to evaluate relaxing constraints included in the model. estat scoretests does not report a joint test. You cannot sum the χ^2 values to obtain a joint test statistic. Thus we learn only that the individual constraints should not be relaxed at reasonable confidence levels.
4. Thus when evaluating multiple constraints, it is better to fit the model without the constraints and use test to evaluate them jointly.

Also see

[SEM] **example 7** — Nonrecursive structural model

[SEM] **sem** — Structural equation model estimation command

[SEM] **sem path notation** — Command syntax for path diagrams

[SEM] **test** — Wald test of linear hypotheses

[SEM] **estat scoretests** — Score tests

Title

example 9 — Structural model with measurement component

Description

To demonstrate a structural model with a measurement component, we use data from Wheaton et al. (1977):

```
. use http://www.stata-press.com/data/r12/sem_sm2
(Structural model with measurement component)

. ssd describe

  Summary statistics data from
  http://www.stata-press.com/data/r12/sem_sm2.dta
    obs:             932                  Structural model with measurem..
   vars:              13                  25 May 2011 11:45
                                          (_dta has notes)
  ------------------------------------------------------------------------
  variable name                  variable label
  ------------------------------------------------------------------------
  educ66                         Education, 1966
  occstat66                      Occupational status, 1966
  anomia66                       Anomia, 1966
  pwless66                       Powerlessness, 1966
  socdist66                      Latin American social distance, 1966
  occstat67                      Occupational status, 1967
  anomia67                       Anomia, 1967
  pwless67                       Powerlessness, 1967
  socdist67                      Latin American social distance, 1967
  occstat71                      Occupational status, 1971
  anomia71                       Anomia, 1971
  pwless71                       Powerlessness, 1971
  socdist71                      Latin American social distance, 1971
  ------------------------------------------------------------------------

. notes

_dta:
  1.  Summary statistics data from Wheaton, B., Muthen B., Alwin, D., &
      Summers, G., 1977, "Assessing reliability and stability in panel models",
      in D. R. Heise (Ed.), _Sociological Methodology 1977_ (pp. 84-136), San
      Francisco: Jossey-Bass, Inc.
  2.  Four indicators each measured in 1966, 1967, and 1981, plus another
      indicator (educ66) measured only in 1966.
  3.  Intended use: Create structural model relating Alienation in 1971,
      Alienation in 1967, and SES in 1966.
```

See *Structural models 3: Unobserved inputs, outputs, or both* in [SEM] **intro 4** for background.

Remarks

Remarks are presented under the following headings:

Fitting the model
Evaluating omitted paths using estat mindices
Refitting the model

Fitting the model

Simplified versions of the model fit by the authors of the referenced paper appear in many SEM software manuals. A simplified model is

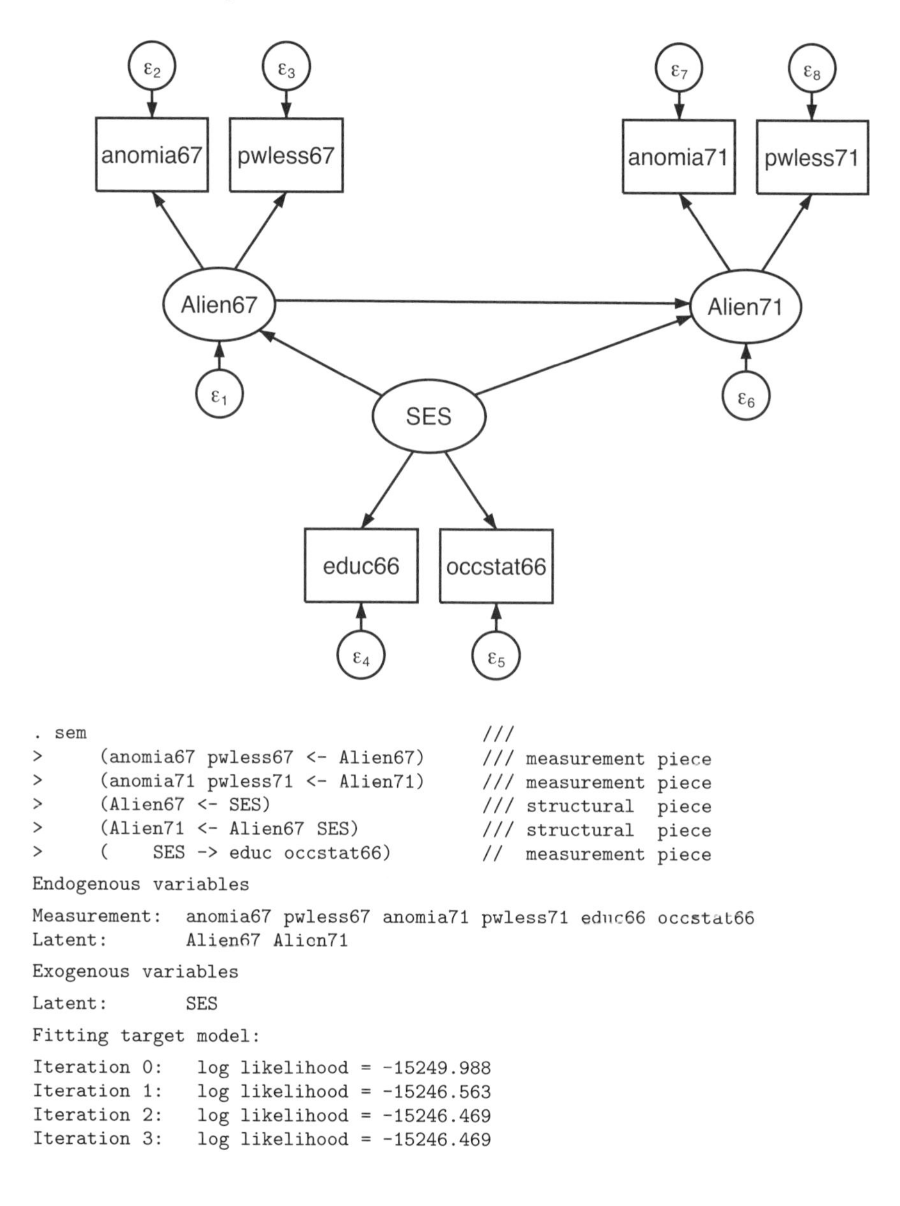


```
. sem                                          ///
>     (anomia67 pwless67 <- Alien67)           /// measurement piece
>     (anomia71 pwless71 <- Alien71)           /// measurement piece
>     (Alien67 <- SES)                         /// structural  piece
>     (Alien71 <- Alien67 SES)                 /// structural  piece
>     (    SES -> educ occstat66)              //  measurement piece
Endogenous variables

Measurement:  anomia67 pwless67 anomia71 pwless71 educ66 occstat66
Latent:       Alien67 Alien71

Exogenous variables

Latent:       SES

Fitting target model:

Iteration 0:   log likelihood = -15249.988
Iteration 1:   log likelihood = -15246.563
Iteration 2:   log likelihood = -15246.469
Iteration 3:   log likelihood = -15246.469
```

```
Structural equation model                       Number of obs      =       932
Estimation method  = ml
Log likelihood     = -15246.469

 ( 1)  [anomia67]Alien67 = 1
 ( 2)  [anomia71]Alien71 = 1
 ( 3)  [educ66]SES = 1
```

	Coef.	OIM Std. Err.	z	P>|z|	[95% Conf.	Interval]
Structural						
Alien71 <-						
Alien67	.7046345	.0533512	13.21	0.000	.6000681	.8092008
SES	-.1744151	.0542400	-3.22	0.001	.280741	-.0680891
Alien67 <-						
SES	-.6140404	.0562407	-10.92	0.000	-.7242701	-.5038107
Measurement						
anom~67 <-						
Alien67	1	(constrained)				
_cons	13.61	.1126205	120.85	0.000	13.38927	13.83073
pwle~67 <-						
Alien67	.8884887	.0431565	20.59	0.000	.8039034	.9730739
_cons	14.67	.1001798	146.44	0.000	14.47365	14.86635
anom~71 <-						
Alien71	1	(constrained)				
_cons	14.13	.1158943	121.92	0.000	13.90285	14.35715
pwle~71 <-						
Alien71	.848602	.0415205	20.44	0.000	.7672233	.9299806
_cons	14.9	.1034537	144.03	0.000	14.69723	15.10277
educ66 <-						
SES	1	(constrained)				
_cons	10.9	.1014894	107.40	0.000	10.70108	11.09892
occs~66 <-						
SES	5.331259	.4307503	12.38	0.000	4.487004	6.175514
_cons	37.49	.6947112	53.96	0.000	36.12839	38.85161
Variance						
e.anomia67	4.00992	.3582979			3.365724	4.777416
e.pwless67	3.187469	.2833741			2.677762	3.794197
e.anomia71	3.695589	.3911515			3.00324	4.547547
e.pwless71	3.621533	.3037911			3.072485	4.268696
e.educ66	2.943819	.5002527			2.109909	4.10732
e.occst~66	260.63	18.24573			227.2139	298.9605
e.Alien67	5.301416	.4831441			4.434225	6.338201
e.Alien71	3.737291	.3881554			3.048955	4.581026
SES	6.65587	.6409483			5.511066	8.038481

```
LR test of model vs. saturated: chi2(6)   =     71.62, Prob > chi2 = 0.0000
```

Notes:

1. Measurement component: In both 1967 and 1971, anomia and powerlessness are used to measure endogenous latent variables representing alienation for the same two years. Education and occupational status are used to measure the exogenous latent variable SES.

2. Structural component: `SES->Alien67` and `SES->Alien71`, and `Alien67->Alien71`.
3. The model versus saturated χ^2 test indicates that the model is a poor fit.

Evaluating omitted paths using estat mindices

That the model is a poor fit leads us to look at the modification indices:

```
. estat mindices
Modification indices
```

	MI	df	P>MI	EPC	Standard EPC
Measurement					
anomia67 <-					
anomia71	51.977	1	0.00	.3906429	.4019988
pwless71	32.517	1	0.00	-.2969288	-.2727601
educ66	5.627	1	0.02	.0935049	.0842631
pwless67 <-					
anomia71	41.618	1	0.00	-.3106997	-.3594369
pwless71	23.622	1	0.00	.2249714	.2323234
educ66	6.441	1	0.01	-.0889042	-.0900664
anomia71 <-					
anomia67	58.768	1	0.00	.4294368	.4173058
pwless67	38.142	1	0.00	-.3873074	-.3347911
pwless71 <-					
anomia67	46.188	1	0.00	-.330848	-.3601637
pwless67	27.761	1	0.00	.2871715	.2780838
educ66 <-					
anomia67	4.415	1	0.04	.1055965	.1171781
pwless67	6.816	1	0.01	-.1469373	-.1450413
Covariance					
e.anomia67					
e.anomia71	63.786	1	0.00	1.951578	.506963
e.pwless71	49.892	1	0.00	-1.506703	-.395379
e.educ66	6.063	1	0.01	.5527616	.1608846
e.pwless67					
e.anomia71	49.876	1	0.00	-1.5342	-.4470098
e.pwless71	37.358	1	0.00	1.159125	.3411622
e.educ66	7.752	1	0.01	-.5557802	-.1814365

EPC = expected parameter change

Notes:

1. There are lots of statistically significant paths we could add to the model.
2. Some of those statistically significant paths also make theoretical sense.
3. Two in particular that make theoretical sense are the covariances between `e.anomia67` and `e.anomia71` and between `e.pwless67` and `e.pwless71`.

Refitting the model

Let's refit the model and include those two previously excluded covariances:

```
. sem                                              ///
>     (anomia67 pwless67 <- Alien67)               /// measurement piece
>     (anomia71 pwless71 <- Alien71)               /// measurement piece
>     (Alien67 <- SES)                             /// structural  piece
>     (Alien71 <- Alien67 SES)                     /// structural  piece
>     (    SES -> educ occstat66)                  /// measurement piece
>                      , cov(e.anomia67*e.anomia71) ///
>                        cov(e.pwless67*e.pwless71)

Endogenous variables

Measurement:  anomia67 pwless67 anomia71 pwless71 educ66 occstat66
Latent:       Alien67 Alien71

Exogenous variables

Latent:       SES

Fitting target model:

Iteration 0:   log likelihood = -15249.988
Iteration 1:   log likelihood = -15217.939
Iteration 2:   log likelihood = -15213.131
Iteration 3:   log likelihood = -15213.046
Iteration 4:   log likelihood = -15213.046
```

```
Structural equation model                       Number of obs      =       932
Estimation method  = ml
Log likelihood     = -15213.046

 ( 1)  [anomia67]Alien67 = 1
 ( 2)  [anomia71]Alien71 = 1
 ( 3)  [educ66]SES = 1

------------------------------------------------------------------------------
             |                 OIM
             |      Coef.   Std. Err.      z    P>|z|     [95% Conf. Interval]
-------------+----------------------------------------------------------------
Structural   |
  Alien71 <- |
     Alien67 |    .606954   .0512305    11.85   0.000      .506544    .7073641
         SES |  -.2270302   .0530773    -4.28   0.000    -.3310598   -.1230006
  -----------+----------------------------------------------------------------
  Alien67 <- |
         SES |   -.575223    .057961    -9.92   0.000    -.6888245   -.4616214
-------------+----------------------------------------------------------------
Measurement  |
  anom~67 <- |
     Alien67 |          1  (constrained)
       _cons |      13.61   .1126143   120.85   0.000     13.38928    13.83072
  -----------+----------------------------------------------------------------
  pwle~67 <- |
     Alien67 |   .9785951   .0619825    15.79   0.000     .8571117    1.100079
       _cons |      14.67   .1001814   146.43   0.000     14.47365    14.86635
  -----------+----------------------------------------------------------------
  anom~71 <- |
     Alien71 |          1  (constrained)
       _cons |      14.13   .1159036   121.91   0.000     13.90283    14.35717
  -----------+----------------------------------------------------------------
  pwle~71 <- |
     Alien71 |   .9217508   .0597225    15.43   0.000     .8046968    1.038805
       _cons |       14.9   .1034517   144.03   0.000     14.69724    15.10276
  -----------+----------------------------------------------------------------
  educ66 <-  |
         SES |          1  (constrained)
       _cons |       10.9   .1014894   107.40   0.000     10.70108    11.09892
  -----------+----------------------------------------------------------------
  occs~66 <- |
         SES |    5.22132    .425595    12.27   0.000     4.387169     6.05547
       _cons |      37.49   .6947112    53.96   0.000     36.12839    38.85161
-------------+----------------------------------------------------------------
Variance     |
  e.anomia67 |   4.728874   .4562989                      3.914024    5.713365
  e.pwless67 |   2.563413   .4060731                      1.879225      3.4967
  e.anomia71 |   4.396081   .5171154                      3.490905    5.535966
  e.pwless71 |   3.072085   .4360331                      2.326049    4.057398
    e.educ66 |   2.803674   .5115854                       1.96069    4.009091
  e.occst~66 |   264.5311   18.22483                      231.1178    302.7751
   e.Alien67 |   4.842058   .4622536                      4.015771    5.838363
   e.Alien71 |   4.084249   .4038993                      3.364614    4.957802
         SES |   6.796014   .6524867                      5.630283    8.203106
-------------+----------------------------------------------------------------
Covariance   |
  e.anomia67 |
  e.anomia71 |   1.622024   .3154266     5.14   0.000       1.0038    2.240249
  -----------+----------------------------------------------------------------
  e.pwless67 |
  e.pwless71 |   .3399961    .262754     1.29   0.196    -.1749923    .8549846
------------------------------------------------------------------------------
LR test of model vs. saturated: chi2(4)   =      4.78, Prob > chi2 = 0.3111
```

Notes:

1. We find the covariance between `e.anomia67` and `e.anomia71` to be significant ($Z = 5.14$).
2. We find the covariance between `e.pwless67` and `e.pwless71` to be insignificant at the 5% level ($Z = 1.29$).
3. The model versus saturated χ^2 test indicates that the model is a good fit.

Also see

[SEM] **estat mindices** — Modification indices

[SEM] **test** — Wald test of linear hypotheses

Title

example 10 — MIMIC model

Description

To demonstrate a MIMIC model, we use the following summary statistics data:

```
. use http://www.stata-press.com/data/r12/sem_mimic1
(Multiple indicators and multiple causes)

. ssd describe

  Summary statistics data from
  http://www.stata-press.com/data/r12/sem_mimic1.dta
    obs:              432                Multiple indicators and multip..
   vars:                5                25 May 2011 10:13
                                         (_dta has notes)

  variable name                  variable label

  occpres                        occupational prestige, two-digit Dunca..
  income                         total family income in units of $2000,..
  s_occpres                      subjective occupational prestige
  s_income                       subjective income
  s_socstat                      subjective overall social status

. notes

_dta:
  1.  Summary statistics data from Kluegel, J. R., R. Singleton, Jr., and C. E.
      Starnes, 1977, "Subjective class identification:  A multiple indicator
      approach", _American Sociological Review_, 42: 599-611.
  2.  Data is also analyzed in Bollen, K. A., 1989, _Structural Equations
      with Latent Variables_, New York: John Wiley & Sons, Inc.
  3.  The summary statistics represent 432 white adults included
      in the sample for the 1969 Gary Area Project for the Institute of Social
      Research at Indiana University.
  4.  The three subjective variables are measures of socioeconomic status based
      on an individuals perception of their own income, occupational prestige,
      and social status.
  5.  The income and occpres variables are objective measures of income and
      occupational prestige, respectively.
```

See *Structural models 4: MIMIC* in [SEM] **intro 4** for background.

Remarks

Remarks are presented under the following headings:

Fitting the MIMIC model
Evaluating the residuals using estat residuals
Performing likelihood-ratio tests using lrtest

Fitting the MIMIC model

Based on the data referenced above, Bollen (1989, 397–399) fits a MIMIC model, the path diagram of which is

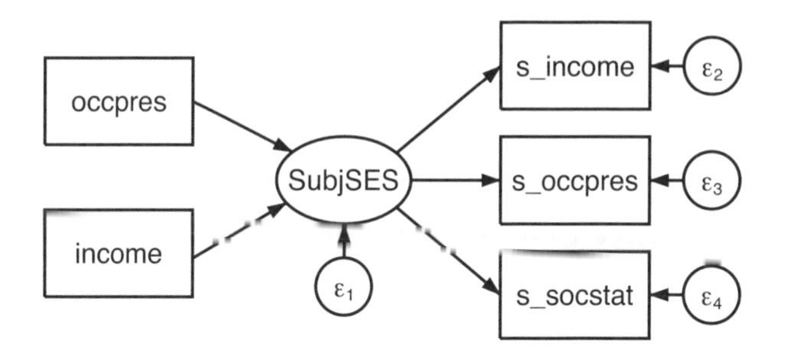

In Bollen (1989, 397–399), he includes paths that he constrains and we do not show. Our model is nonetheless equivalent to the one he shows. In his textbook, Bollen illustrates various ways the same model can be written.

```
. sem (SubjSES -> s_income s_occpres s_socstat) (SubjSES <- income occpres)
Endogenous variables

Measurement:  s_income s_occpres s_socstat
Latent:       SubjSES

Exogenous variables

Observed:     income occpres

Fitting target model:

Iteration 0:   log likelihood = -4252.1834  (not concave)
Iteration 1:   log likelihood = -4022.9057  (not concave)
Iteration 2:   log likelihood =   -3994.24
Iteration 3:   log likelihood = -3978.5284  (not concave)
Iteration 4:   log likelihood = -3974.5499
Iteration 5:   log likelihood = -3973.1229
Iteration 6:   log likelihood = -3971.9427
Iteration 7:   log likelihood = -3971.9236
Iteration 8:   log likelihood = -3971.9236

Structural equation model                       Number of obs      =       432
Estimation method  = ml
Log likelihood     = -3971.9236

 ( 1)  [s_income]SubjSES = 1

------------------------------------------------------------------------------
             |                 OIM
             |      Coef.   Std. Err.      z    P>|z|     [95% Conf. Interval]
-------------+----------------------------------------------------------------
Structural   |
  SubjSES <- |
      income |   .0827327   .0138499     5.97   0.000     .0555874     .109878
     occpres |   .0046275   .0012464     3.71   0.000     .0021847    .0070704
-------------+----------------------------------------------------------------
Measurement  |
  s_inc~e <- |
     SubjSES |          1  (constrained)
       _cons |   .9612057   .0794155    12.10   0.000     .8055541    1.116857
  -----------+----------------------------------------------------------------
  s_occ~s <- |
     SubjSES |   .7301313   .0832913     8.77   0.000     .5668832    .8933793
       _cons |   1.114563   .0656195    16.99   0.000     .9859513    1.243175
  -----------+----------------------------------------------------------------
  s_soc~t <- |
     SubjSES |   .9405104   .0934852    10.06   0.000     .7572827    1.123738
       _cons |   1.002114   .0706576    14.18   0.000     .8636274      1.1406
-------------+----------------------------------------------------------------
Variance     |
  e.s_income |   .2087534   .0254099                       .164446    .2649987
  e.s_occp~s |   .2811856   .0228914                      .2397156    .3298296
  e.s_socs~t |    .180714   .0218405                      .1425996    .2290157
   e.SubjSES |    .186012    .027048                      .1398838    .2473513
------------------------------------------------------------------------------
LR test of model vs. saturated: chi2(4)   =     26.65, Prob > chi2 = 0.0000
```

Notes:

1. In this model, there are three observed variables that record the person's idea of their perceived socioeconomic status (SES). One is the person's general idea of their SES (`s_socstat`); another is based on their income (`s_income`); and the last is based on their occupational prestige (`s_occpres`). Those three variables form the latent variable `SubjSES`.

2. The other two observed variables are the person's income (`income`) and occupation, the latter measured by the two-digit Duncan SEI scores for occupations (`occpres`). These two variables are treated as predictors of `SubjSES`.

3. In the model, (1) is viewed as subjective and (2) is viewed as objective.
4. All variables are statistically significant at the 5% level, but the model versus saturated test suggests that we are not modeling the covariances well.

Evaluating the residuals using estat residuals

Remember that SEM fits covariances and means. Residuals in the SEM sense thus refer to covariances and means. If we are not fitting well, we can examine the residuals.

```
. estat residuals, normalized
Residuals of observed variables
  Mean residuals
```

	s_income	s_occpres	s_socstat	income	occpres
raw	0.000	0.000	0.000	0.000	0.000
normalized	0.000	0.000	0.000	0.000	0.000

```
  Covariance residuals
```

	s_income	s_occpres	s_socstat	income	occpres
s_income	-0.000				
s_occpres	-0.009	0.000			
s_socstat	0.000	0.008	0.000		
income	0.101	-0.079	-0.053	0.000	
occpres	-0.856	1.482	0.049	0.000	0.000

```
  Normalized covariance residuals
```

	s_income	s_occpres	s_socstat	income	occpres
s_income	-0.000				
s_occpres	-0.425	0.000			
s_socstat	0.008	0.401	0.000		
income	1.362	-1.137	-0.771	0.000	
occpres	-1.221	2.234	0.074	0.000	0.000

Notes:

1. The residuals can be partitioned into two subsets: mean residuals and covariance residuals.
2. The `normalized` option caused the normalized residuals to be displayed.
3. Concerning mean residuals, the raw residuals and the normalized residuals are shown on a separate line of the first table.
4. Concerning covariance residuals, the raw residuals and the normalized residuals are shown in separate tables.
5. Distinguish between normalized residuals and standardized residuals. Both are available from `estat residuals`; if we wanted standardized residuals, we would have specified the `standardized` option instead of or along with `normalized`.
6. Both normalized and standardized residuals attempt to adjust the residuals in the same way. The normalized residuals are always valid, but they do not follow a standard normal distribution. The standardized residuals do follow a standard normal distribution if they can be calculated; otherwise, they will equal missing values. When both can be calculated (equivalent to both being appropriate), the normalized residuals will be a little smaller than the standardized residuals.

7. The normalized covariance residuals between income and s_income and between occpres and s_occpres are large.

Performing likelihood-ratio tests using lrtest

Thus Bollen suggests adding a direct path from the objective measures to the corresponding subjective measures. We are about to fit the model

```
(SubjSES -> s_income s_occpres s_socstat)    ///
(SubjSES <- income occpres)                  ///
(s_income <- income)                         ///  <- new
(s_occpres <- occpres)                       //   <- new
```

For no other reason than we want to demonstrate the likelihood-ratio test, we will then use lrtest rather than test to test the joint significance of the new paths. lrtest compares the likelihood values of two fitted models. Thus we will use lrtest to compare this new model with the one above. To do that, we must plan ahead and store in memory the currently fit model:

```
. estimates store mimic1
```

Alternatively, we could skip that and calculate the joint significance of the two new paths using a Wald test and the test command.

In any case, having stored the current estimates under the name mimic1, we can now fit our new model:

```
. sem (SubjSES -> s_income s_occpres s_socstat)
>      (SubjSES <- income occpres)
>     (s_income <- income)
>     (s_occpres <- occpres)
Endogenous variables

Observed:     s_income s_occpres
Measurement:  s_socstat
Latent:       SubjSES

Exogenous variables

Observed:     income occpres

Fitting target model:

Iteration 0:   log likelihood = -4267.0974  (not concave)
Iteration 1:   log likelihood = -4022.8637  (not concave)
Iteration 2:   log likelihood = -3977.1937
Iteration 3:   log likelihood = -3962.9248
Iteration 4:   log likelihood = -3961.5382
Iteration 5:   log likelihood = -3960.7634
Iteration 6:   log likelihood = -3960.7112
Iteration 7:   log likelihood = -3960.7111
```

```
Structural equation model                       Number of obs      =       432
Estimation method  = ml
Log likelihood     = -3960.7111
 ( 1)  [s_income]SubjSES = 1
```

	Coef.	OIM Std. Err.	z	P>|z|	[95% Conf.	Interval]
Structural						
s_inc~e <-						
SubjSES	1	(constrained)				
income	.0532426	.0142861	3.73	0.000	.0252423	.081243
_cons	.8825316	.0781684	11.29	0.000	.7293243	1.035739
s_occ~s <-						
SubjSES	.7837824	.1011453	7.75	0.000	.5855412	.9820235
occpres	.0045201	.0013552	3.34	0.001	.0018641	.0071762
_cons	1.06586	.0696057	15.31	0.000	.9294357	1.202285
SubjSES <-						
income	.0538023	.0129157	4.17	0.000	.0284881	.0791166
occpres	.0034324	.0011217	3.06	0.002	.0012339	.0056309
Measurement						
s_soc~t <-						
SubjSES	1.195539	.158271	7.55	0.000	.8853336	1.505745
_cons	1.07922	.0783231	13.78	0.000	.9257099	1.232731
Variance						
e.s_income	.22927	.0248903			.1853267	.2836327
e.s_occp~s	.2773785	.0223972			.2367782	.3249405
e.s_socs~t	.1459008	.0282278			.0998559	.2131777
e.SubjSES	.1480268	.0278376			.1023919	.2140007

```
LR test of model vs. saturated: chi2(2)   =      4.22, Prob > chi2 = 0.1211
```

Now we can perform the likelihood-ratio test:

```
. lrtest mimic1 .

Likelihood-ratio test                                 LR chi2(2)  =     22.42
(Assumption: mimic1 nested in .)                      Prob > chi2 =    0.0000
```

Notes:

1. The syntax of `lrtest` is `lrtest` *modelname1 modelname2*. We specified the first model name as `mimic1`, the model we previously stored. We specified the second model name as period (`.`), meaning the model most recently fit. The order in which we specify the names is irrelevant.
2. We find the two added paths to be whoppingly significant.

Also see

[SEM] **sem** — Structural equation model estimation command

[SEM] **estat residuals** — Display mean and covariance residuals

[SEM] **lrtest** — Likelihood-ratio test of linear hypothesis

Title

example 11 — estat framework

Description

To demonstrate `estat framework`, which displays results in Bentler–Weeks form, we continue from where [SEM] **example 10** left off:

```
. use http://www.stata-press.com/data/r12/sem_mimic1
. ssd describe
. notes
. sem (SubjSES -> s_income s_occpres s_socstat)    ///
      (SubjSES <- income occpres)
. estat residuals, normalized
. estimates store mimic1
. sem (SubjSES -> s_income s_occpres s_socstat)    ///
      (SubjSES <- income occpres)                  ///
      (s_income <- income)                         ///
      (s_occpres <- occpres)
. lrtest mimic1 .
```

See *Structural models 4: MIMIC* in [SEM] **intro 4** for background.

Remarks

If you prefer to see SEM results reported in Bentler–Weeks form, type `estat framework` after estimating using `sem`. Many people find Bentler–Weeks form helpful in understanding how the model is fit.

[SEM] **example 10** ended by fitting

```
. sem (SubjSES   -> s_income s_occpres s_socstat)    ///
      (SubjSES   <- income occpres)                  ///
      (s_income  <- income)                          ///
      (s_occpres <- occpres)
```

In Bentler–Weeks form, the output appears as

```
. estat framework, fitted
Endogenous variables on endogenous variables
```

Beta	observed s_income	s_occpres	s_socstat	latent SubjSES
observed				
s_income	0	0	0	1
s_occpres	0	0	0	.7837824
s_socstat	0	0	0	1.195539
latent				
SubjSES	0	0	0	0

Exogenous variables on endogenous variables

Gamma	observed income	occpres
observed		
s_income	.0532426	0
s_occpres	0	.0045201
s_socstat	0	0
latent		
SubjSES	.0538023	.0034324

Covariances of error variables

Psi	observed e.s_inc~e	e.s_occ~s	e.s_soc~t	latent e.SubjSES
observed				
e.s_income	.22927			
e.s_occpres	0	.2773785		
e.s_socstat	0	0	.1459008	
latent				
e.SubjSES	0	0	0	.1480268

Intercepts of endogenous variables

alpha	observed s_income	s_occpres	s_socstat	latent SubjSES
_cons	.8825316	1.06586	1.07922	0

Covariances of exogenous variables

Phi	observed income	occpres
observed		
income	4.820021	
occpres	13.62431	451.6628

Means of exogenous variables

kappa	observed income	occpres
mean	5.04	36.698

Fitted covariances of observed and latent variables

Sigma	observed s_income	s_occpres	s_socstat	latent SubjSES	observed income
observed					
s_income	.4478605				
s_occpres	.1614442	.4086515			
s_socstat	.2255141	.1738218	.3922179		
latent					
SubjSES	.1886296	.1453919	.2060302	.1723325	
observed					
income	.5627229	.3014933	.3659453	.3060923	4.820021
occpres	3.008694	3.831182	2.729774	2.2833	13.62431

Sigma	observed occpres
observed	
occpres	451.6628

Fitted means of observed and latent variables

mu	observed s_income	s_occpres	s_socstat	latent SubjSES	observed income
mu	1.548	1.543	1.553999	.3971255	5.04

mu	observed occpres
mu	36.698

Notes:

1. Bentler–Weeks form is a vector and matrix notation for the estimated parameters of the model. The matrices are known as β, Γ, Ψ, α, Φ, and κ. Those Greek names are spelled out in the labels, along with a header stating what each contains.
2. We specified `estat framework` option `fitted`. That caused `estat framework` to list one more matrix and one more vector at the end: Σ and μ. These two results are especially interesting to those wishing to see the ingredients of the residuals reported by `estat residuals`.
3. One of the more useful results reported by `estat framework, fitted` is the Σ matrix, which reports all estimated covariances in a readable format and includes the model-implied covariances that do not appear in `sem`'s ordinary output.
4. `estat framework` also allows the `standardized` option if you want standardized output.

Also see

[SEM] **example 10** — MIMIC model

[SEM] **estat framework** — Display estimation results in modeling framework

Title

example 12 — Seemingly unrelated regression

Description

sem can be used to estimate seemingly unrelated regression. We will use auto.dta, which surely needs no introduction:

```
. sysuse auto
(1978 Automobile Data)
```

See *Structural models 5: Seemingly unrelated regression (SUR)* in [SEM] **intro 4**.

Remarks

We fit the following model:

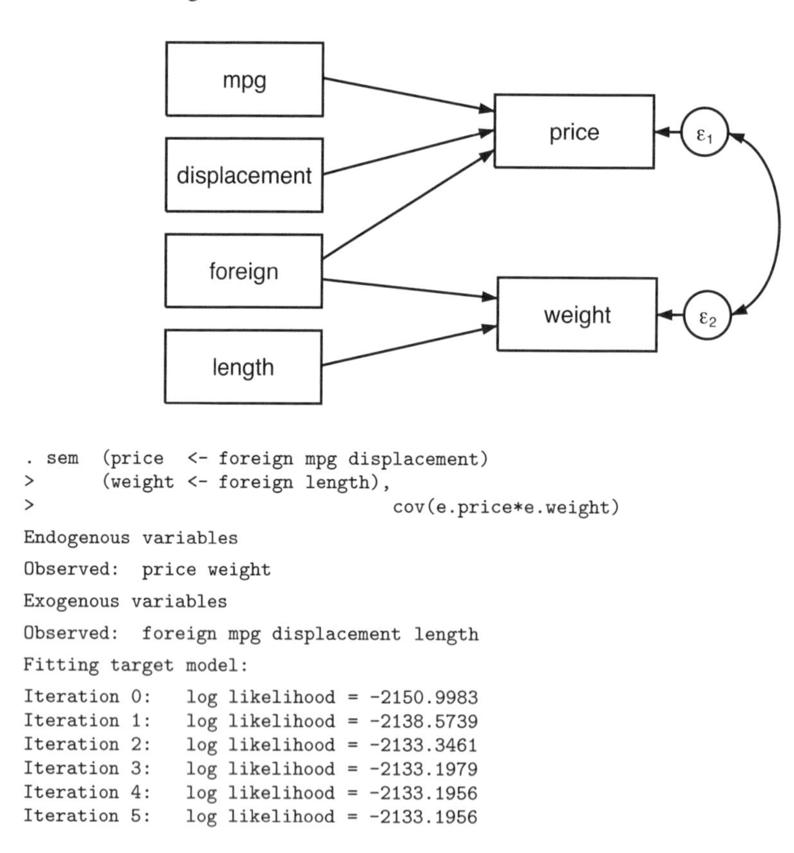


```
. sem  (price  <- foreign mpg displacement)
>      (weight <- foreign length),
>                                   cov(e.price*e.weight)
Endogenous variables
Observed:  price weight
Exogenous variables
Observed:  foreign mpg displacement length
Fitting target model:
Iteration 0:   log likelihood = -2150.9983
Iteration 1:   log likelihood = -2138.5739
Iteration 2:   log likelihood = -2133.3461
Iteration 3:   log likelihood = -2133.1979
Iteration 4:   log likelihood = -2133.1956
Iteration 5:   log likelihood = -2133.1956
```

```
Structural equation model                       Number of obs      =         74
Estimation method  = ml
Log likelihood     = -2133.1956

------------------------------------------------------------------------------
             |                 OIM
             |      Coef.   Std. Err.      z    P>|z|     [95% Conf. Interval]
-------------+----------------------------------------------------------------
Structural   |
  price <-   |
     foreign |   2940.929   724.7311     4.06   0.000     1520.482    4361.376
         mpg |  -105.0163   57.93461    -1.81   0.070     -218.566     8.53347
  displace~t |   17.22083     4.5941     3.75   0.000     8.216558     26.2251
       _cons |   4129.866   1984.253     2.08   0.037     240.8022    8018.931
  -----------+----------------------------------------------------------------
  weight <-  |
     foreign |  -153.2515   76.21732    -2.01   0.044    -302.6347   -3.868275
      length |   30.73507   1.584743    19.39   0.000     27.62903    33.84111
       _cons |  -2711.096   312.6813    -8.67   0.000     -3323.94   -2098.252
-------------+----------------------------------------------------------------
Variance     |
     e.price |    4732491   801783.1                       3395302     6596312
    e.weight |   60253.09   9933.316                      43616.45    83235.44
-------------+----------------------------------------------------------------
Covariance   |
  e.price    |
    e.weight |     209268   73909.54     2.83   0.005     64407.92      354128
------------------------------------------------------------------------------
LR test of model vs. saturated: chi2(3)   =     38.86, Prob > chi2 = 0.0000
```

Notes:

1. Point estimates are the same as reported by

```
. sureg (price foreign mpg displ) (weight foreign length), isure
```

   sureg's isure option is required to make sureg iterate to the maximum likelihood estimate.

2. If you wish to compare the estimated variances and covariances after estimation by sureg, type

```
. matrix list e(Sigma)
```

   sureg does not estimate standard errors on variances and covariances.

3. Standard errors will be different between sem and sureg. In this case, there is no reason to prefer one set of standard errors over the other, and standard errors are asymptotically equivalent. This is a case of exogenous variables only on the right-hand side. When the model being fit is recursive, standard errors produced by sem are better than those from sureg, both asymptotically and in finite samples.

4. One reason you might want to use sem is that sem will provide robust standard errors whereas sureg does not.

5. Multivariate regression can be viewed as seemingly unrelated regression. You just need to specify the same regressors for each equation. In that case, standard errors reported by sem will be the same as those reported by mvreg if one applies the multiplicative $\sqrt{(N-p-1)/N}$ degree-of-freedom adjustment.

Also see

[SEM] **example 13** — Equation-level Wald test

[SEM] **sem** — Structural equation model estimation command

Title

example 13 — Equation-level Wald test

Description

This example picks up where [SEM] **example 12** left off:

```
. use http://www.stata-press.com/data/r12/auto
. sem  (price  <- foreign mpg displacement)            ///
       (weight <- foreign length),                     ///
                   cov(e.price*e.weight)
```

We demonstrate `estat eqtest`. See [SEM] **intro 6** and see [SEM] **estat eqtest**.

Remarks

We have fit a two-equation model with equations for endogenous variables `price` and `weight`. There happen to be two equations, the model happens to be a seemingly unrelated regression, and the endogenous variables happen to be observed, but none of that is important right now.

`estat eqtest` displays equation-by-equation Wald tests that all coefficients excluding the intercepts are 0.

```
. estat eqtest
Wald tests for equations
```

	chi2	df	p
observed			
price	36.43	3	0.0000
weight	633.34	2	0.0000

Notes:

1. The null hypothesis for this test is that the coefficients other than the intercepts are 0. We can reject that null hypothesis for each equation.

Also see

[SEM] **example 12** — Seemingly unrelated regression

[SEM] **intro 6** — Postestimation tests and predictions

[SEM] **estat eqtest** — Equation-level test that all coefficients are zero

Title

example 14 — Predicted values

Description

This example picks up where the first part of [SEM] **example 1** left off:

```
. use http://www.stata-press.com/data/r12/sem_1fmm
. sem (x1 x2 x3 x4 <- X)
```

We demonstrate the use of `predict`. See [SEM] **intro 6** and see [SEM] **predict**.

Remarks

`predict` can create new variables containing predicted values of (1) observed endogenous variables, (2) latent variables, whether endogenous or exogenous, and (3) latent endogenous variables. In the case of latent variables, item (2) corresponds to the factor score and item (3) is the linear prediction.

Below we demonstrate (1) and (2):

```
. predict x1hat x2hat, xb(x1 x2)
. predict Xhat, latent(X)
```

You specify options on `predict` to specify what you want predicted and how. Because of the differing options, the two commands could not have been combined into one command.

Our dataset now contains three new variables. Below we compare the three variables with the original `x1` and `x2` by using first `summarize` and then `correlate`:

```
. summarize x1 x1hat x2 x2hat Xhat
```

Variable	Obs	Mean	Std. Dev.	Min	Max
x1	123	96.28455	14.16444	54	131
x1hat	123	96.28455	10.65716	68.42469	122.9454
x2	123	97.28455	16.14764	64	135
x2hat	123	97.28455	12.49406	64.62267	128.5408
Xhat	123	-1.66e-08	10.65716	-27.85986	26.66084

Notes:

1. Means of `x1hat` and `x1` are identical; means of `x2hat` and `x2` are identical.
2. Standard deviation of `x1hat` is less than that of `x1`; standard deviation of `x2hat` is less than that of `x2`. Some of the variation in `x1` and `x2` is not explained by the model.
3. Standard deviations of `x1hat` and `Xhat` are equal. This is because in

$$x1 = b_0 + b_1 X + e_1$$

coefficient b_1 was constrained to be equal to 1 because of the anchoring normalization constraint; see *Identification 2: Normalization constraints (anchoring)* in [SEM] **intro 3**.

4. The mean of `Xhat` is −1.66e–08 rather than 0. Had we typed

   ```
   . predict double Xhat, latent(X)
   ```

   the mean would have been −1.61e–15.

```
. correlate x1 x1hat x2 x2hat Xhat
(obs=123)

             |       x1    x1hat       x2    x2hat     Xhat
-------------+---------------------------------------------
          x1 |   1.0000
       x1hat |   0.7895   1.0000
          x2 |   0.5826   0.8119   1.0000
       x2hat |   0.7895   1.0000   0.8119   1.0000
        Xhat |   0.7895   1.0000   0.8119   1.0000   1.0000
```

Notes:

1. Both `x1hat` and `x2hat` correlate 1 with `Xhat`. That is because both are linear functions of `Xhat` alone.
2. That `x1hat` and `x2hat` correlate 1 is implied by (1), directly above.
3. That `Xhat`, `x1hat`, and `x2hat` all have the same correlation with `x1` and with `x2` is also implied by (1), directly above.

Also see

[SEM] **example 1** — Single-factor measurement model

[SEM] **intro 6** — Postestimation tests and predictions

[SEM] **predict** — Factor scores, linear predictions, etc.

Title

example 15 — Higher-order CFA

Description

sem can be used to estimate higher-order confirmatory factor analysis models.

```
. use http://www.stata-press.com/data/r12/sem_hcfa1
(Higher-order CFA)

. ssd describe

  Summary statistics data from
  http://www.stata-press.com/data/r12/sem_hcfa1.dta
    obs:              251                  Higher-order CFA
   vars:               16                  25 May 2011 11:26
                                           (_dta has notes)

  variable name                  variable label

  phyab1                         Physical ability 1
  phyab2                         Physical ability 2
  phyab3                         Physical ability 3
  phyab4                         Physical ability 4
  appear1                        Appearance 1
  appear2                        Appearance 2
  appear3                        Appearance 3
  appear4                        Appearance 4
  peerrel1                       Relationship w/ peers 1
  peerrel2                       Relationship w/ peers 2
  peerrel3                       Relationship w/ peers 3
  peerrel4                       Relationship w/ peers 4
  parrel1                        Relationship w/ parent 1
  parrel2                        Relationship w/ parent 2
  parrel3                        Relationship w/ parent 3
  parrel4                        Relationship w/ parent 4

. notes

_dta:
  1.  Summary statistics data from Marsh, H. W. and Hocevar, D., 1985,
      "Application of confirmatory factor analysis to the study of
      self-concept: First- and higher order factor models and their invariance
      across groups", _Psychological Bulletin_, 97: 562-582.
  2.  Summary statistics based on 251 students from Sydney, Australia in Grade
      5.
  3.  Data collected using the Self-Description Questionnaire and includes
      sixteen subscales designed to measure nonacademic traits: four intended
      to measure physical ability, four intended to measure physical
      appearance, four intended to measure relations with peers, and four
      intended to measure relations with parents.
```

See *Higher-order CFA models* in [SEM] **intro 4** for background.

Remarks

We fit the following model:

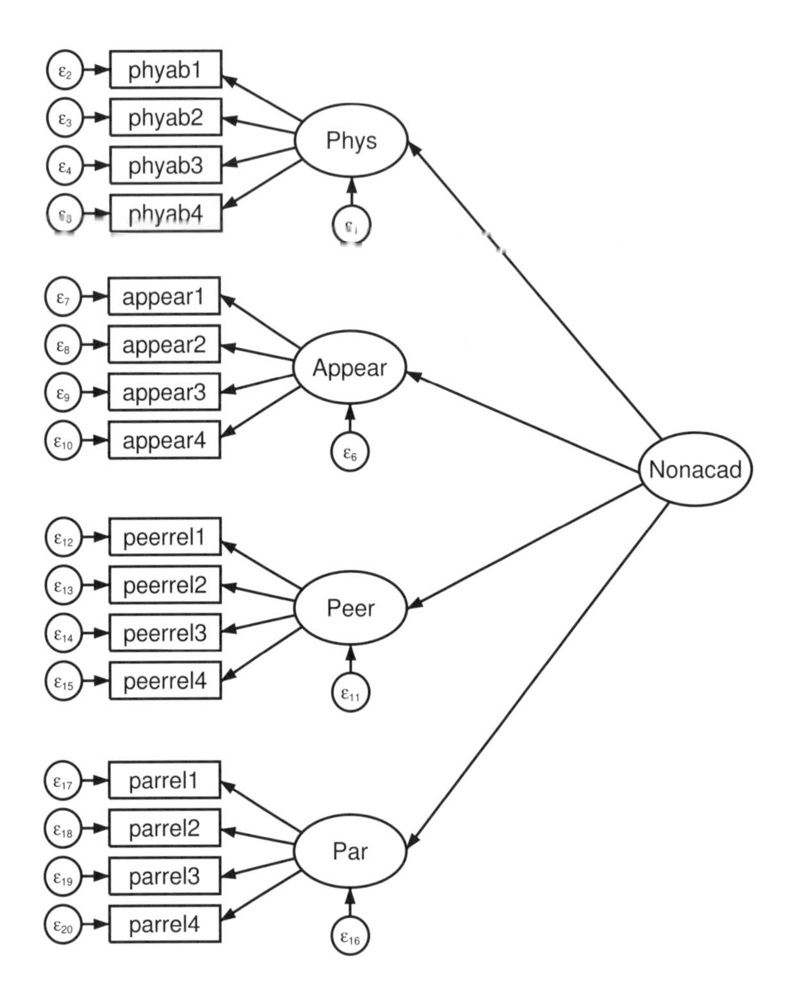

```
. sem (Phys -> phyab1 phyab2 phyab3 phyab4)
>     (Appear -> appear1 appear2 appear3 appear4)
>     (Peer -> peerrel1 peerrel2 peerrel3 peerrel4)
>     (Par -> parrel1 parrel2 parrel3 parrel4)
>     (Nonacad -> Phys Appear Peer Par)
Endogenous variables

Measurement:  phyab1 phyab2 phyab3 phyab4 appear1 appear2 appear3 appear4
              peerrel1 peerrel2 peerrel3 peerrel4 parrel1 parrel2 parrel3
              parrel4
Latent:       Phys Appear Peer Par

Exogenous variables

Latent:       Nonacad

Fitting target model:

Iteration 0:   log likelihood = -7686.6699  (not concave)
Iteration 1:   log likelihood = -7643.7387  (not concave)
Iteration 2:   log likelihood = -7616.2966  (not concave)
Iteration 3:   log likelihood = -7597.6133
Iteration 4:   log likelihood = -7588.9515
Iteration 5:   log likelihood = -7585.3162
Iteration 6:   log likelihood = -7584.8125
Iteration 7:   log likelihood = -7584.7885
Iteration 8:   log likelihood = -7584.7881

Structural equation model                       Number of obs      =       251
Estimation method  = ml
Log likelihood     = -7584.7881

 ( 1)  [phyab1]Phys = 1
 ( 2)  [appear1]Appear = 1
 ( 3)  [peerrel1]Peer = 1
 ( 4)  [parrel1]Par = 1
 ( 5)  [Phys]Nonacad = 1

------------------------------------------------------------------------------
             |                 OIM
             |      Coef.   Std. Err.      z    P>|z|     [95% Conf. Interval]
-------------+----------------------------------------------------------------
Structural   |
  Phys <-    |
     Nonacad |          1  (constrained)
  -----------+----------------------------------------------------------------
  Appear <-  |
     Nonacad |   2.202491   .3975476     5.54   0.000     1.423312     2.98167
  -----------+----------------------------------------------------------------
  Peer <-    |
     Nonacad |   1.448035   .2921383     4.96   0.000     .8754549    2.020616
  -----------+----------------------------------------------------------------
  Par <-     |
     Nonacad |    .569956   .1382741     4.12   0.000     .2989437    .8409683
-------------+----------------------------------------------------------------
Measurement  |
  phyab1 <-  |
        Phys |          1  (constrained)
       _cons |        8.2   .1159065    70.75   0.000     7.972827    8.427173
  -----------+----------------------------------------------------------------
  phyab2 <-  |
        Phys |   .9332477   .1285726     7.26   0.000       .68125    1.185245
       _cons |       8.23    .122207    67.34   0.000     7.990479    8.469521
  -----------+----------------------------------------------------------------
  phyab3 <-  |
        Phys |   1.529936   .1573845     9.72   0.000     1.221468    1.838404
       _cons |       8.17   .1303953    62.66   0.000      7.91443     8.42557
  -----------+----------------------------------------------------------------
```

```
   phyab4 <-
        Phys     1.325641   .1338053     9.91   0.000     1.063387    1.587894
       _cons         8.56   .1146471    74.66   0.000     8.335296    8.784704
  ---------------------------------------------------------------------------
  appear1 <-
      Appear            1  (constrained)
       _cons         7.41   .1474041    50.27   0.000     7.121093    7.698907
  ---------------------------------------------------------------------------
  appear2 <-
      Appear       1.0719   .0821893    13.04   0.000     .9108121    1.232988
       _cons            7   .1644123    42.58   0.000     6.677758    7.322242
  ---------------------------------------------------------------------------
  appear3 <-
      Appear     1.035198   .0893075    11.59   0.000     .8601581    1.210237
       _cons         7.17   .1562231    45.90   0.000     6.863808    7.476192
  ---------------------------------------------------------------------------
  appear4 <-
      Appear     .9424492   .0860848    10.95   0.000     .7737262    1.111172
       _cons          7.4   .1474041    50.20   0.000     7.111093    7.688907
  ---------------------------------------------------------------------------
  peerr~1 <-
        Peer            1  (constrained)
       _cons         8.81   .1077186    81.79   0.000     8.598875    9.021125
  ---------------------------------------------------------------------------
  peerr~2 <-
        Peer     1.214379   .1556051     7.80   0.000     .9093989     1.51936
       _cons         7.94   .1215769    65.31   0.000     7.701714    8.178286
  ---------------------------------------------------------------------------
  peerr~3 <-
        Peer     1.667829    .190761     8.74   0.000     1.293944    2.041714
       _cons         7.52   .1373248    54.76   0.000     7.250848    7.789152
  ---------------------------------------------------------------------------
  peerr~4 <-
        Peer     1.363627    .159982     8.52   0.000     1.050068    1.677186
       _cons         8.29   .1222066    67.84   0.000     8.050479    8.529521
  ---------------------------------------------------------------------------
  parrel1 <-
         Par            1  (constrained)
       _cons         9.35   .0825215   113.30   0.000     9.188261    9.511739
  ---------------------------------------------------------------------------
  parrel2 <-
         Par     1.159754    .184581     6.28   0.000     .7979822    1.521527
       _cons         9.13   .0988998    92.32   0.000      8.93616     9.32384
  ---------------------------------------------------------------------------
  parrel3 <-
         Par     2.035143   .2623826     7.76   0.000     1.520882    2.549403
       _cons         8.67   .1114983    77.76   0.000     8.451467    8.888533
  ---------------------------------------------------------------------------
  parrel4 <-
         Par     1.651802   .2116151     7.81   0.000     1.237044     2.06656
       _cons            9   .0926003    97.19   0.000     8.818507    9.181493
-----------------------------------------------------------------------------
Variance
    e.phyab1      2.07466   .2075636                      1.705244    2.524103
    e.phyab2     2.618638    .252693                      2.167386    3.163841
    e.phyab3     1.231013   .2062531                      .8864333     1.70954
    e.phyab4     1.019261   .1600644                      .7492262    1.386621
   e.appear1     1.986955   .2711164                      1.520699    2.596169
   e.appear2     2.801673   .3526427                      2.189162    3.585561
   e.appear3      2.41072    .300262                      1.888545    3.077276
   e.appear4     2.374508   .2872554                      1.873267    3.009868
   e.peerrel1    1.866632     .18965                      1.529595    2.277933
```

```
   e.peerrel2 |   2.167766   .2288099                      1.762654    2.665984
   e.peerrel3 |   1.824346   .2516762                      1.392131    2.390749
   e.peerrel4 |   1.803918    .212599                      1.431856    2.272659
    e.parrel1 |   1.214141   .1195921                      1.000982    1.472692
    e.parrel2 |   1.789125   .1748043                      1.477322    2.166738
    e.parrel3 |   1.069717   .1767086                      .7738511    1.478702
    e.parrel4 |   .8013735    .121231                      .5957527    1.077963
       e.Phys |    .911538   .1933432                      .6014913    1.381403
     e.Appear |    1.59518   .3704939                      1.011838    2.514828
       e.Peer |   .2368108   .1193956                      .0881539    .6361528
        e.Par |   .3697854   .0915049                      .2276755     .600597
      Nonacad |   .3858166   .1237638                      .2057449    .7234903
-------------------------------------------------------------------------------
LR test of model vs. saturated: chi2(100) =    219.48, Prob > chi2 = 0.0000
```

Notes:

1. The idea behind this model is that physical ability, appearance, and relationships with peers and parents may be determined by a latent variable containing nonacademic traits. This model was suggested by Bollen (1989, 315).
2. `sem` automatically provided normalization constraints for the first-order factors `Phys`, `Appear`, `Peer`, and `Par`. Their path coefficients were set to 1.
3. `sem` automatically provided a normalization constraint for the second-order factor `Nonacad`. Its path coefficient was set to 1.

Also see

[SEM] **sem** — Structural equation model estimation command

Title

example 16 — Correlation

Description

sem can be used to produce correlations or covariances between exogenous variables. The advantages of using sem over Stata's correlate command are (1) you can perform statistical tests on the results and (2) you can handle missing values in a more elegant way.

To demonstrate these features, we use

```
. use http://www.stata-press.com/data/r12/census13
(1980 Census data by state)
. describe
Contains data from http://www.stata-press.com/data/r12/census13.dta
  obs:            50                          1980 Census data by state
 vars:             9                          9 Apr 2011 10:09
 size:         1,600

              storage  display     value
variable name   type   format      label      variable label

state           long   %13.0g      state1     State
brate           long   %10.0g                 Birth rate
pop             long   %12.0gc                Population
medage          float  %9.2f                  Median age
division        int    %8.0g       division   Census Division
region          int    %-8.0g      cenreg     Census region
mrgrate         float  %9.0g
dvcrate         float  %9.0g
medagesq        float  %9.0g

Sorted by:
```

See *Correlations* in [SEM] **intro 4** for background.

Remarks

Remarks are presented under the following headings:

Using sem to obtain correlation matrices

We fit the following model:

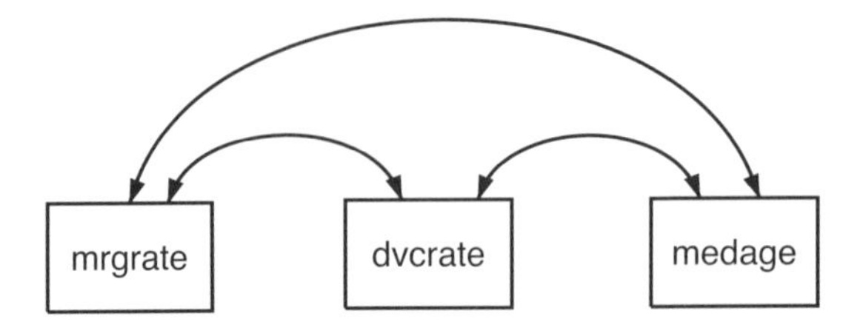

This model does nothing more than estimate the covariances (correlations), something we could obtain from the `correlate` command by typing

```
. correlate mrgrate dvcrate medage
(obs=50)

             |  mrgrate  dvcrate   medage
-------------+---------------------------
     mrgrate |   1.0000
     dvcrate |   0.7700   1.0000
      medage |  -0.0177  -0.2229   1.0000

. correlate mrgrate dvcrate medage, covariance
(obs=50)

             |  mrgrate  dvcrate   medage
-------------+---------------------------
     mrgrate |  .000662
     dvcrate |  .000063  1.0e-05
      medage | -.000769 -.001191  2.86775
```

As explained in *Correlations* in [SEM] **intro 4**, to see results presented as correlations rather than as covariances, we specify `sem`'s `standardized` option:

```
. sem ( <- mrgrate dvcrate medage), standardized

Exogenous variables

Observed:  mrgrate dvcrate medage

Fitting target model:

Iteration 0:   log likelihood =  258.58985
Iteration 1:   log likelihood =  258.58985

Structural equation model                       Number of obs      =        50
Estimation method  = ml
Log likelihood     =  258.58985

------------------------------------------------------------------------------
             |                 OIM
Standardized |      Coef.   Std. Err.      z    P>|z|     [95% Conf. Interval]
-------------+----------------------------------------------------------------
Mean         |
     mrgrate |   .7332509   .1593002     4.60   0.000     .4210282    1.045474
     dvcrate |   2.553791    .291922     8.75   0.000     1.981634    3.125947
      medage |   17.62083   1.767749     9.97   0.000     14.15611    21.08556
-------------+----------------------------------------------------------------
Variance     |
     mrgrate |          1          .                             .           .
     dvcrate |          1          .                             .           .
      medage |          1          .                             .           .
-------------+----------------------------------------------------------------
Covariance   |
  mrgrate    |
     dvcrate |   .7699637   .0575805    13.37   0.000     .6571079    .8828195
      medage |  -.0176541   .1413773    -0.12   0.901    -.2947485    .2594403
  -----------+----------------------------------------------------------------
  dvcrate    |
      medage |   -.222932   .1343929    -1.66   0.097    -.4863373    .0404732
------------------------------------------------------------------------------
LR test of model vs. saturated: chi2(0)   =      0.00, Prob > chi2 =      .
```

Notes:

1. The correlations reported are

	sem	correlate
mrgrate and dvcrate	0.7699637	0.7700
mrgrate and medage	−0.0176541	−0.0177
dvcrate and medage	−0.222932	−0.2229

Testing correlations using estat stdize and test

We can test whether the correlations between median age and marriage and divorce rates are equal using `test`, by typing

```
. estat stdize:   ///
        test _b[cov(medage,mrgrate):_cons] = _b[cov(medage,dvcrate):_cons]
```

We must prefix `test` with `estat stdize` because otherwise we would be testing equality of covariances; see *Displaying other results, statistics, and tests* in [SEM] **intro 6** and see [SEM] **estat stdize**.

That we refer to the two correlations (covariances) by typing `_b[cov(medage,mrgrate):_cons]` and `_b[cov(medage,dvcrate):_cons]` is something nobody remembers and that we remind ourselves of by redisplaying `sem` results with the `coeflegend` option:

```
. sem, coeflegend

Structural equation model                       Number of obs      =        50
Estimation method  = ml
Log likelihood     =  258.58985

------------------------------------------------------------------------------
             |      Coef.  Legend
-------------+----------------------------------------------------------------
Mean         |
     mrgrate |   .0186789  _b[mean(mrgrate):_cons]
     dvcrate |   .0079769  _b[mean(dvcrate):_cons]
      medage |      29.54  _b[mean(medage):_cons]
-------------+----------------------------------------------------------------
Variance     |
     mrgrate |   .0006489  _b[var(mrgrate):_cons]
     dvcrate |   9.76e-06  _b[var(dvcrate):_cons]
      medage |     2.8104  _b[var(medage):_cons]
-------------+----------------------------------------------------------------
Covariance   |
  mrgrate    |
     dvcrate |   .0000613  _b[cov(mrgrate,dvcrate):_cons]
      medage |  -.0007539  _b[cov(mrgrate,medage):_cons]
  -----------+----------------------------------------------------------------
  dvcrate    |
      medage |  -.0011674  _b[cov(dvcrate,medage):_cons]
------------------------------------------------------------------------------
LR test of model vs. saturated: chi2(0)   =      0.00, Prob > chi2 =      .
```

We can now obtain the test:

```
. estat stdize:
>         test _b[cov(medage,mrgrate):_cons] = _b[cov(medage,dvcrate):_cons]
 ( 1)  [cov(mrgrate,medage)]_cons - [cov(dvcrate,medage)]_cons = 0
           chi2(  1) =    4.78
         Prob > chi2 =    0.0288
```

Notes:

1. We can reject the test at the 5% level.

Also see

[SEM] **test** — Wald test of linear hypotheses

[SEM] **estat stdize** — Test standardized parameters

[R] **correlate** — Correlations (covariances) of variables or coefficients

Title

example 17 — Correlated uniqueness model

Description

To demonstrate a correlated uniqueness model, we use the following summary statistics data:

```
. use http://www.stata-press.com/data/r12/sem_cu1
(Correlated uniqueness)

. ssd describe

  Summary statistics data from
  http://www.stata-press.com/data/r12/sem_cu1.dta
    obs:           500                 Correlated uniqueness
   vars:             9                 25 May 2011 10:12
                                       (_dta has notes)

  variable name                 variable label

  par_i                         self-report inventory for paranoid
  szt_i                         self-report inventory for schizotypal
  szd_i                         self-report inventory for schizoid
  par_c                         clinical interview rating for paranoid
  szt_c                         clinical interview rating for schizoty..
  szd_c                         clinical interview rating for schizoid
  par_o                         observer rating for paranoid
  szt_o                         observer rating for schizotypal
  szd_o                         observer rating for schizoid

. notes

_dta:
  1.  Summary statistics data for Multitrait-Multimethod matrix (a specific
      kind of correlation matrix) and standard deviations from Brown, Timothy
      A., 2006, _Confirmatory Factor Analysis for Applied Research_, New York,
      NY: The Guilford Press.
  2.  Summary statistics represent a sample of 500 patients who were evaluated
      for three personality disorders using three different methods.
  3.  The personality disorders include paranoid, schizotypal, and schizoid.
  4.  The methods of evaluation include a self-report inventory, ratings from a
      clinical interview, and observational ratings.
```

See *Correlated uniqueness model* in [SEM] **intro 4** for background.

Remarks

We fit the following model:

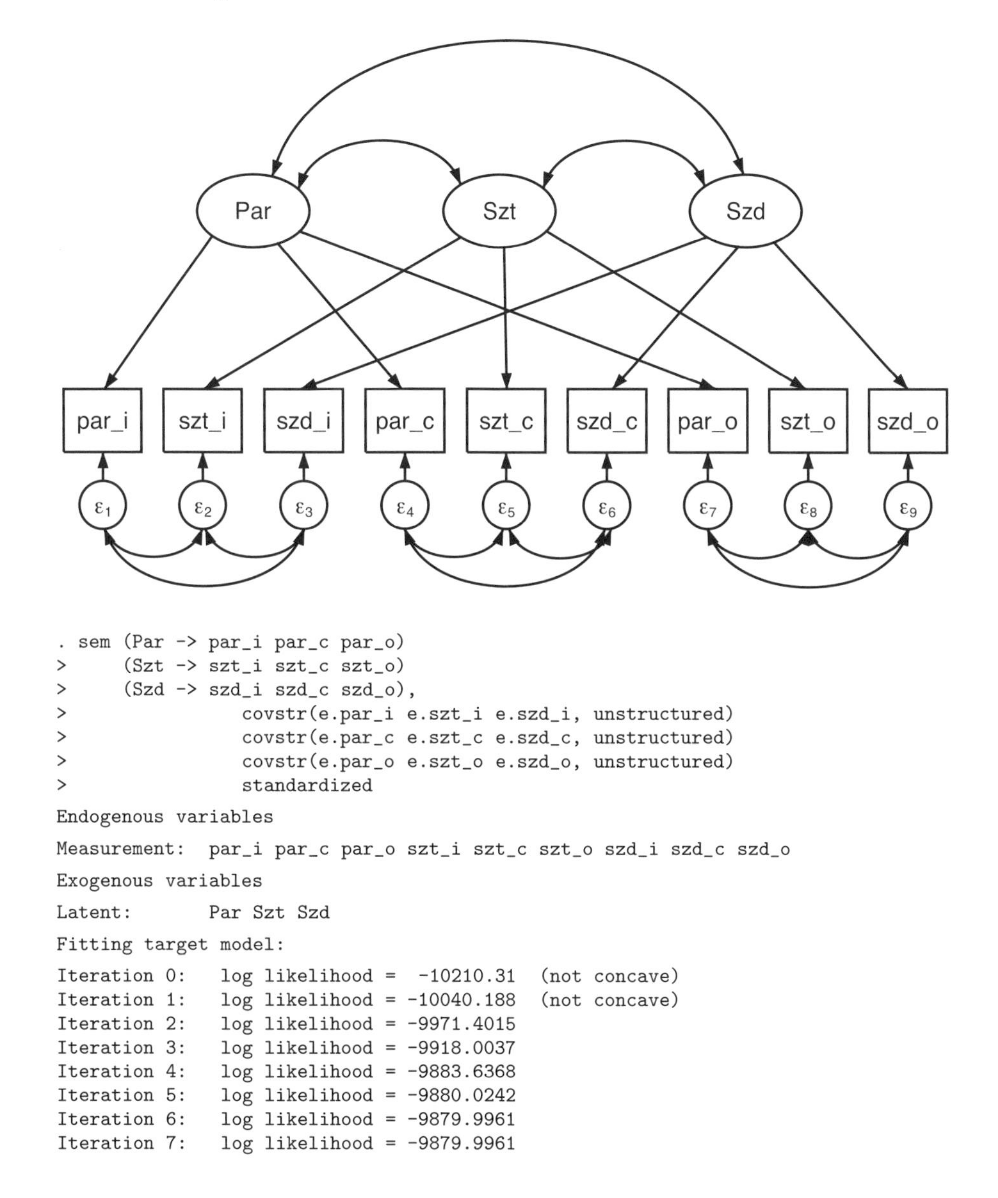


```
. sem (Par -> par_i par_c par_o)
>     (Szt -> szt_i szt_c szt_o)
>     (Szd -> szd_i szd_c szd_o),
>              covstr(e.par_i e.szt_i e.szd_i, unstructured)
>              covstr(e.par_c e.szt_c e.szd_c, unstructured)
>              covstr(e.par_o e.szt_o e.szd_o, unstructured)
>              standardized
Endogenous variables

Measurement:  par_i par_c par_o szt_i szt_c szt_o szd_i szd_c szd_o

Exogenous variables

Latent:       Par Szt Szd

Fitting target model:

Iteration 0:   log likelihood =  -10210.31  (not concave)
Iteration 1:   log likelihood = -10040.188  (not concave)
Iteration 2:   log likelihood = -9971.4015
Iteration 3:   log likelihood = -9918.0037
Iteration 4:   log likelihood = -9883.6368
Iteration 5:   log likelihood = -9880.0242
Iteration 6:   log likelihood = -9879.9961
Iteration 7:   log likelihood = -9879.9961
```

```
Structural equation model                       Number of obs      =       500
Estimation method  = ml
Log likelihood     = -9879.9961

 ( 1)  [par_i]Par = 1
 ( 2)  [szt_i]Szt = 1
 ( 3)  [szd_i]Szd = 1

------------------------------------------------------------------------------
             |                 OIM
Standardized |      Coef.   Std. Err.      z    P>|z|     [95% Conf. Interval]
-------------+----------------------------------------------------------------
Measurement  |
  par_i <-   |
         Par |   .7119709   .0261858    27.19   0.000     .6606476    .7632941
  -----------+----------------------------------------------------------------
  par_c <-   |
         Par |   .8410183   .0242205    34.72   0.000     .7935469    .8884897
  -----------+----------------------------------------------------------------
  par_o <-   |
         Par |   .7876062   .0237685    33.14   0.000     .7410209    .8341916
  -----------+----------------------------------------------------------------
  szt_i <-   |
         Szt |   .7880887   .0202704    38.88   0.000     .7483594    .8278179
  -----------+----------------------------------------------------------------
  szt_c <-   |
         Szt |   .7675732   .0244004    31.46   0.000     .7197493    .8153972
  -----------+----------------------------------------------------------------
  szt_o <-   |
         Szt |   .8431662   .0181632    46.42   0.000      .807567    .8787653
  -----------+----------------------------------------------------------------
  szd_i <-   |
         Szd |   .7692321   .0196626    39.12   0.000     .7306942      .80777
  -----------+----------------------------------------------------------------
  szd_c <-   |
         Szd |   .8604596   .0179455    47.95   0.000     .8252871    .8956321
  -----------+----------------------------------------------------------------
  szd_o <-   |
         Szd |   .8715597   .0155875    55.91   0.000     .8410086    .9021107
-------------+----------------------------------------------------------------
Variance     |
     e.par_i |   .4930975   .0372871                      .4251739    .5718722
     e.par_c |   .2926882   .0407398                      .2228049    .3844905
     e.par_o |   .3796764   .0374404                      .3129503    .4606295
     e.szt_i |   .3789163   .0319498                      .3211966    .4470082
     e.szt_c |   .4108313   .0374582                      .3436006    .4912169
     e.szt_o |   .2890708   .0306291                      .2348623    .3557912
     e.szd_i |    .408282   .0302501                      .3530966    .4720922
     e.szd_c |   .2596093   .0308827                      .2056187    .3277766
     e.szd_o |   .2403837    .027171                       .192616    .2999976
         Par |          1          .                             .           .
         Szt |          1          .                             .           .
         Szd |          1          .                             .           .
-------------+----------------------------------------------------------------
Covariance   |
  e.par_i    |
     e.szt_i |   .2166732   .0535966     4.04   0.000     .1116258    .3217207
     e.szd_i |   .4411039   .0451782     9.76   0.000     .3525563    .5296515
  -----------+----------------------------------------------------------------
  e.par_c    |
     e.szt_c |  -.1074802   .0691107    -1.56   0.120    -.2429348    .0279743
     e.szd_c |  -.2646125   .0836965    -3.16   0.002    -.4286546   -.1005705
  -----------+----------------------------------------------------------------
```

e.par_o						
e.szt_o	.4132457	.0571588	7.23	0.000	.3012165	.5252749
e.szd_o	.3684402	.0587572	6.27	0.000	.2532781	.4836022
e.szt_i						
e.szd_i	.7456394	.0351079	21.24	0.000	.6768292	.8144496
e.szt_c						
e.szd_c	-.3296552	.0720069	-4.58	0.000	-.4707861	-.1885244
e.szt_o						
e.szd_o	.4781276	.0588923	8.12	0.000	.3627009	.5935544
Par						
Szt	.3806759	.045698	8.33	0.000	.2911095	.4702422
Szd	.3590146	.0456235	7.87	0.000	.2695941	.4484351
Szt						
Szd	.3103837	.0466126	6.66	0.000	.2190246	.4017428

```
LR test of model vs. saturated: chi2(15)  =     14.37, Prob > chi2 = 0.4976
```

Notes:

1. We use the correlated uniqueness model fit above to analyze a multitrait–multimethod (MTMM) matrix. The MTMM matrix was developed by Campbell and Fiske (1959) to evaluate construct validity of measures. Each trait is measured by using different methods, and the correlation matrix produced is used to evaluate whether measures that are related in theory are related in fact (convergent validity) and whether measures that are not intended to be related are not related in fact (discriminant validity).

   In this example, the traits are the latent variables `Par`, `Sct`, and `Sz`.

   The observed variables are the method–trait combinations.

   The observed traits are the personality disorders paranoid (`par`), schotypal (`szt`), and schizoid (`szd`). The methods used to measure them are self report (`_i`), clinical interview (`_c`), and observer rating (`_o`). Thus variable `par_i` is paranoid (`par`) measured by self-report (`_i`).

2. Note our use of the `covstructure()` option, which we abbreviated to `covstr()`. We used this option instead of `cov()` to save typing; see *Correlated uniqueness model* in [SEM] **intro 4**.

3. Large values of the factor loadings (path coefficients) indicate convergent validity.

4. Small correlations between latent variables indicate discriminant validity.

Also see

[SEM] **sem** — Structural equation model estimation command

[SEM] **sem option covstructure()** — Specifying covariance restrictions

Title

example 18 — Latent growth model

Description

To demonstrate a latent growth model, we use the following data:

```
. use http://www.stata-press.com/data/r12/sem_lcm

. describe

Contains data from http://www.stata-press.com/data/r12/sem_lcm.dta
  obs:           359
 vars:             4                          25 May 2011 11:08
 size:         5,744                          (_dta has notes)
-------------------------------------------------------------------------------
              storage  display     value
variable name   type   format      label      variable label
-------------------------------------------------------------------------------
lncrime0        float  %9.0g                  ln(crime rate) in Jan & Feb
lncrime1        float  %9.0g                  ln(crime rate) in Mar & Apr
lncrime2        float  %9.0g                  ln(crime rate) in May & Jun
lncrime3        float  %9.0g                  ln(crime rate) in Jul & Aug
-------------------------------------------------------------------------------
Sorted by:

. notes

_dta:
  1.  Data used in Bollen, Kenneth A. and Patrick J. Curran, 2006, _Latent
      Curve Models: A Structural Equation Perspective_. Hoboken, New Jersey:
      John Wiley & Sons
  2.  Data from 1995 Uniform Crime Reports for 359 communities in New York
      state.
```

See *Latent growth models* in [SEM] **intro 4** for background.

Remarks

We fit the following model:

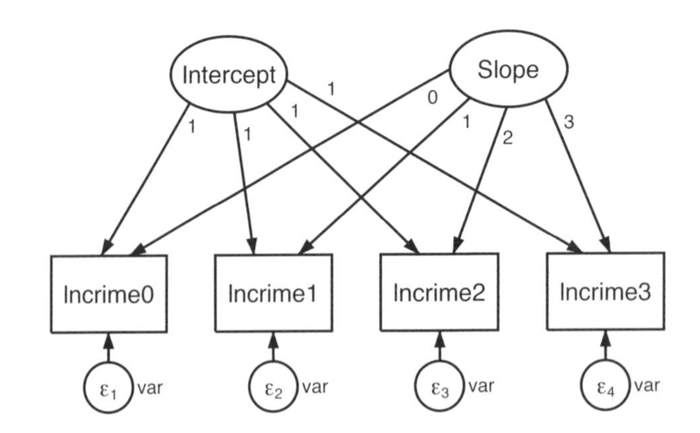

```
. sem (lncrime0 <- Intercept@1 Slope@0 _cons@0)
>     (lncrime1 <- Intercept@1 Slope@1 _cons@0)
>     (lncrime2 <- Intercept@1 Slope@2 _cons@0)
>     (lncrime3 <- Intercept@1 Slope@3 _cons@0),
>             latent(Intercept Slope)
>             var(e.lncrime0@var e.lncrime1@var
>                 e.lncrime2@var e.lncrime3@var)
>             means(Intercept Slope)
Endogenous variables

Measurement:  lncrime0 lncrime1 lncrime2 lncrime3

Exogenous variables

Latent:       Intercept Slope

Fitting target model:

Iteration 0:   log likelihood = -1034.1038
Iteration 1:   log likelihood = -1033.9044
Iteration 2:   log likelihood = -1033.9037
Iteration 3:   log likelihood = -1033.9037

Structural equation model                       Number of obs      =       359
Estimation method  = ml
Log likelihood     = -1033.9037

 ( 1)  [lncrime0]Intercept = 1
 ( 2)  [lncrime1]Intercept = 1
 ( 3)  [lncrime1]Slope = 1
 ( 4)  [lncrime2]Intercept = 1
 ( 5)  [lncrime2]Slope = 2
 ( 6)  [lncrime3]Intercept = 1
 ( 7)  [lncrime3]Slope = 3
 ( 8)  [var(e.lncrime0)]_cons - [var(e.lncrime3)]_cons = 0
 ( 9)  [var(e.lncrime1)]_cons - [var(e.lncrime3)]_cons = 0
 (10)  [var(e.lncrime2)]_cons - [var(e.lncrime3)]_cons = 0
 (11)  [lncrime0]_cons = 0
 (12)  [lncrime1]_cons = 0
 (13)  [lncrime2]_cons = 0
 (14)  [lncrime3]_cons = 0

------------------------------------------------------------------------------
             |                 OIM
             |      Coef.   Std. Err.      z    P>|z|     [95% Conf. Interval]
-------------+----------------------------------------------------------------
Measurement  |
  lncri~0 <- |
   Intercept |          1  (constrained)
       _cons |          0  (constrained)
  -----------+----------------------------------------------------------------
  lncri~1 <- |
   Intercept |          1  (constrained)
       Slope |          1  (constrained)
       _cons |          0  (constrained)
  -----------+----------------------------------------------------------------
  lncri~2 <- |
   Intercept |          1  (constrained)
       Slope |          2  (constrained)
       _cons |          0  (constrained)
  -----------+----------------------------------------------------------------
  lncri~3 <- |
   Intercept |          1  (constrained)
       Slope |          3  (constrained)
       _cons |          0  (constrained)
-------------+----------------------------------------------------------------
```

	Coef.	Std. Err.	z	P>|z|	[95% Conf.	Interval]
Mean						
Intercept	5.337915	.0407501	130.99	0.000	5.258047	5.417784
Slope	.1426952	.0104574	13.65	0.000	.1221992	.1631912
Variance						
e.lncrime0	.0981956	.0051826			.0885457	.1088972
e.lncrime1	.0981956	.0051826			.0885457	.1088972
e.lncrime2	.0981956	.0051826			.0885457	.1088972
e.lncrime3	.0981956	.0051826			.0885457	.1088972
Intercept	.527409	.0446436			.4467822	.6225858
Slope	.0196198	.0031082			.0143829	.0267635
Covariance						
Intercept						
Slope	-.034316	.0088848	-3.86	0.000	-.0517298	-.0169022

LR test of model vs. saturated: chi2(8) = 16.25, Prob > chi2 = 0.0390

Notes:

1. In this example, we have repeated measures of the crime rate in 1995. We will assume that the underlying rate grows linearly.

2. As explained in *Latent growth models* in [SEM] **intro 4**, we assume

$$\texttt{lncrime}_i = \texttt{Intercept} + i \times \texttt{Slope}$$

3. `sem` does not usually report the means of latent exogenous variables because `sem` automatically includes the identifying constraint that the means are 0; see *How sem solves the problem for you* in [SEM] **intro 3** and see *Default normalization constraints* in [SEM] **sem**.

   In this case, `sem` did not constrain the means to be 0 because we specified `sem`'s `means()` option. In particular, we specified `means(Intercept Slope)`, which said not to constrain the means of those two exogenous latent variables and to report the estimated result.

   Our model was identified even without the usual 0 constraints on `Intercept` and `Slope` because we specified enough other constraints.

4. We estimate the `Intercept` to have mean 5.34 and the mean `Slope` to be 0.14 per two-months. Remember, we have measured crime rates as log base e crime rates.

5. It might help some to think of this as a mixed model:

```
. generate id = _n
. reshape long lncrime, i(id) j(year)
. xtmixed lncrime year || id:year, cov(unstructred) mle var
```

   The mean `Intercept` and `Slope` are what `xtmixed` would refer to as the coefficients in the fixed-effects part of the model.

Also see

[SEM] **sem** — Structural equation model estimation command

Title

example 19 — Creating multiple-group summary statistics data

Description

The data analyzed in [SEM] **example 20** are summary statistics data and contain summary statistics on two groups of subjects, those from grade 4 and those from grade 5. Below we show how we created this summary statistics dataset.

See [SEM] **intro 10** for background on summary statistics data.

Remarks

See [SEM] **example 2** for creating a single-group dataset from published covariances. In this example, we will create a two-group dataset from published correlations, standard deviations, and means.

Marsh and Hocevar (1985) publish lots of summary statistics data, of which we will enter the data for students in grade 4 and grade 5 found on pages 579–581. In that source, the authors published the correlations, standard deviations, and means of their variables.

We will (1) set the data for the first group, (2) declare that we have groups and wish to add another, and (3) set the data for the second group.

Starting with the first group, we will issue the commands:

```
. ssd init      variable names
. ssd set obs   values
. ssd set means values
. ssd set sd    values
. ssd set corr  values
```

We will first set the end-of-line delimiter to a semicolon because we are going to have some long lines. We will be entering summary statistics data for 16 variables!

```
. #delimit ;
delimiter now ;
. ssd init phyab1   phyab2   phyab3   phyab4
>          appear1  appear2  appear3  appear4
>          peerrel1 peerrel2 peerrel3 peerrel4
>          parrel1  parrel2  parrel3  parrel4 ;
Summary statistics data initialized.  Next use, in any order,
    ssd set observations (required)
        It is best to do this first.
    ssd set means (optional)
        Default setting is 0.
    ssd set variances or ssd set sd (optional)
        Use this only if you have set or will set correlations and, even
        then, this is optional but highly recommended.  Default setting is 1.
    ssd set covariances or ssd set correlations (required)
```

```
. ssd set obs 134 ;
  (value set)

    Status:
                               observations:    set
                                      means:  unset
                          variances or sd:  unset
               covariances or correlations:  unset (required to be set)

. ssd set means
>      8.34 8.34 8.37 8.40 7.51 7.22 7.03 7.13
>      8.44 7.62 7.06 7.89 9.32 9.39 8.69 9.13 ;
  (values set)

    Status:
                               observations:    set
                                      means:    set
                          variances or sd:  unset
               covariances or correlations:  unset (required to be set)

. ssd set sd
>      1.90 1.75 2.06 1.88 2.30 2.63 2.71 2.42
>      2.05 2.22 2.38 2.12 1.21 1.21 1.71 1.32 ;
  (values set)

    Status:
                               observations:    set
                                      means:    set
                          variances or sd:    set
               covariances or correlations:  unset (required to be set)

. ssd set corr
>      1.0 \
>      .50 1.0 \
>      .59 .46 1.0 \
>      .58 .43 .66 1.0 \
>      .30 .27 .35 .46 1.0 \
>      .32 .34 .38 .39 .71 1.0 \
>      .38 .41 .43 .53 .68 .67 1.0 \
>      .23 .29 .33 .43 .61 .63 .73 1.0 \
>      .43 .32 .40 .42 .36 .34 .45 .42 1.0 \
>      .38 .40 .38 .49 .53 .61 .69 .59 .59 1.0 \
>      .27 .24 .41 .37 .43 .46 .57 .57 .61 .59 1.0 \
>      .43 .41 .37 .47 .51 .45 .63 .61 .59 .58 .65 1.0 \
>      .20 .14 .15 .18 .22 .21 .13 .03 .15 .19 .12 .14 1.0 \
>      .29 .18 .26 .20 .25 .29 .17 .25 .35 .23 .23 .28 .25 1.0 \
>      .37 .14 .34 .37 .34 .34 .35 .33 .42 .36 .39 .39 .53 .50 1.0 \
>      .13 .10 .16 .21 .33 .28 .23 .22 .23 .25 .23 .28 .46 .43 .59 1.0 ;
  (values set)

    Status:
                               observations:    set
                                      means:    set
                          variances or sd:    set
               covariances or correlations:    set

. #delimit cr
delimiter now cr
```

We have now entered the data for the first group, and ssd reports that we have a fully set dataset.

Next we are going to add a second group by typing

```
. ssd addgroup grade
  (new group grade==2 added)
    The ssd set commands now modify the new group grade==2.  If you need to
    modify data for grade==1, place a 1 right after the set.  For example,
        . ssd set 1 means ...
    would modify the means for group grade==1.
```

The `ssd set` command now modifies the new group `grade==2`. If we needed to modify data for `grade==1`, we would place a 1 right after the `set`. For example,

```
. ssd set 1 means ...
```

We are not modifying data; however, we are now adding data for the second group. The procedure for entering the second group is the same as the procedure for entering the first group:

```
. ssd set obs   values
. ssd set means values
. ssd set sd    values
. ssd set corr  values
```

We do that below.

```
. #delimit ;
delimiter now ;
. ssd set obs 251 ;
  (value set for group grade==2)
    Status for group grade==2:
                        observations:    set
                               means:  unset
                   variances or sd:  unset
       covariances or correlations:  unset (required to be set)
. ssd set corr
>      1.0 \
>      .31 1.0 \
>      .52 .45 1.0 \
>      .54 .46 .70 1.0 \
>      .15 .33 .22 .21 1.0 \
>      .14 .28 .21 .13 .72 1.0 \
>      .16 .32 .35 .31 .59 .56 1.0 \
>      .23 .29 .43 .36 .55 .51 .65 1.0 \
>      .24 .13 .24 .23 .25 .24 .24 .30 1.0 \
>      .19 .26 .22 .18 .34 .37 .36 .32 .38 1.0 \
>      .16 .24 .36 .30 .33 .29 .44 .51 .47 .50 1.0 \
>      .16 .21 .35 .24 .31 .33 .41 .39 .47 .47 .55 1.0 \
>      .08 .18 .09 .12 .19 .24 .08 .21 .21 .19 .19 .20 1.0 \
>      .01 -.01 .03 .02 .10 .13 .03 .05 .26 .17 .23 .26 .33 1.0 \
>      .06 .19 .22 .22 .23 .24 .20 .26 .16 .23 .38 .24 .42 .40 1.0 \
>      .04 .17 .10 .07 .26 .24 .12 .26 .16 .22 .32 .17 .42 .42 .65 1.0 ;
  (values set for group grade==2)
    Status for group grade==2:
                        observations:    set
                               means:  unset
                   variances or sd:  unset
       covariances or correlations:    set
```

```
. ssd set sd      1.84 1.94 2.07 1.82 2.34 2.61 2.48 2.34
>                 1.71 1.93 2.18 1.94 1.31 1.57 1.77 1.47 ;
  (values set for group grade==2)

    Status for group grade==2:
                       observations:    set
                              means:  unset
                   variances or sd:     set
       covariances or correlations:     set

. ssd set means  8.20 8.23 8.17 8.56 7.41 7.00 7.17 7.40
>                8.81 7.94 7.52 8.29 9.35 9.13 8.67 9.00 ;
  (values set for group grade==2)

    Status for group grade==2:
                       observations:    set
                              means:    set
                   variances or sd:     set
       covariances or correlations:     set

. #delimit cr
delimiter now cr
```

We could stop here and save the data in a Stata dataset. We might type

```
. save sem_2fmmby
```

However, we intend to use this data as an example in this manual and online. Here is what you would see if you typed `ssd describe`:

```
. ssd describe

  Summary statistics data
    obs:              385
   vars:               16
  ------------------------------------------------------------------------

  variable name                 variable label
  ------------------------------------------------------------------------

  phyab1
  phyab2
  phyab3
  phyab4
  appear1
  appear2
  appear3
  appear4
  peerrel1
  peerrel2
  peerrel3
  peerrel4
  parrel1
  parrel2
  parrel3
  parrel4
  ------------------------------------------------------------------------

  Group variable:  grade  (2 groups)
   Obs. by group:  134, 251
```

We are going to label these data so that `ssd describe` can provide more information:

```
. label data "two-factor CFA"
. label var phyab1   "Physical ability 1"
. label var phyab2   "Physical ability 2"
. label var phyab3   "Physical ability 3"
. label var phyab4   "Physical ability 4"
. label var appear1   "Appearance 1"
. label var appear2   "Appearance 2"
. label var appear3   "Appearance 3"
. label var appear4   "Appearance 4"
. label var peerrel1  "Relationship w/ peers 1"
. label var peerrel2  "Relationship w/ peers 2"
. label var peerrel3  "Relationship w/ peers 3"
. label var peerrel4  "Relationship w/ peers 4"
. label var parrel1   "Relationship w/ parent 1"
. label var parrel2   "Relationship w/ parent 2"
. label var parrel3   "Relationship w/ parent 3"
. label var parrel4   "Relationship w/ parent 4"
. #delimit ;
delimiter now ;
. notes: Summary statistics data from
>        Marsh, H. W. and Hocevar, D., 1985,
>        "Application of confirmatory factor analysis to the study of
>        self-concept: First- and higher order factor models and their
>        invariance across groups", _Psychological Bulletin_, 97: 562-582. ;
. notes: Summary statistics based on
>        134 students in grade 4 and
>        251 students in grade 5
>        from Sydney, Australia. ;
. notes: Group 1 is grade 4, group 2 is grade 5. ;
. notes: Data collected using the Self-Description Questionnaire
>        and includes sixteen subscales designed to measure
>        nonacademic traits:  four intended to measure physical
>        ability, four intended to measure physical appearance,
>        four intended to measure relations with peers, and four
>        intended to measure relations with parents. ;
. #delimit cr
delimiter now cr
```

We would now save the dataset.

To see `ssd describe`'s output with the data labeled, see [SEM] **example 20**.

Also see

[SEM] **ssd** — Making summary statistics data

[SEM] **example 20** — Two-factor measurement model by group

Title

example 20 — Two-factor measurement model by group

Description

Below we demonstrate sem's group() option, which allows fitting models in which path coefficients and covariances differ across groups of the data, such as for males and females. We use the following data:

```
. use http://www.stata-press.com/data/r12/sem_2fmmby
(two-factor CFA)

. ssd describe

  Summary statistics data from
  http://www.stata-press.com/data/r12/sem_2fmmby.dta
    obs:          385                  two-factor CFA
   vars:           16                  25 May 2011 11:11
                                       (_dta has notes)
  ------------------------------------------------------------------------
  variable name                 variable label
  ------------------------------------------------------------------------
  phyab1                        Physical ability 1
  phyab2                        Physical ability 2
  phyab3                        Physical ability 3
  phyab4                        Physical ability 4
  appear1                       Appearance 1
  appear2                       Appearance 2
  appear3                       Appearance 3
  appear4                       Appearance 4
  peerrel1                      Relationship w/ peers 1
  peerrel2                      Relationship w/ peers 2
  peerrel3                      Relationship w/ peers 3
  peerrel4                      Relationship w/ peers 4
  parrel1                       Relationship w/ parent 1
  parrel2                       Relationship w/ parent 2
  parrel3                       Relationship w/ parent 3
  parrel4                       Relationship w/ parent 4
  ------------------------------------------------------------------------
  Group variable:  grade  (2 groups)
   Obs. by group:  134, 251

. notes

_dta:
  1.  Summary statistics data from Marsh, H. W. and Hocevar, D., 1985,
      "Application of confirmatory factor analysis to the study of
      self-concept: First- and higher order factor models and their invariance
      across groups", _Psychological Bulletin_, 97: 562-582.
  2.  Summary statistics based on 134 students in grade 4 and 251 students in
      grade 5 from Sydney, Australia.
  3.  Group 1 is grade 4, group 2 is grade 5.
  4.  Data collected using the Self-Description Questionnaire and includes
      sixteen subscales designed to measure nonacademic status: four intended
      to measure physical ability, four intended to measure physical
      appearance, four intended to measure relations with peers, and four
      intended to measure relations with parents.
```

Remarks

Remarks are presented under the following headings:

Background

See [SEM] **intro 5** for background on sem's group() option.

We will fit the model

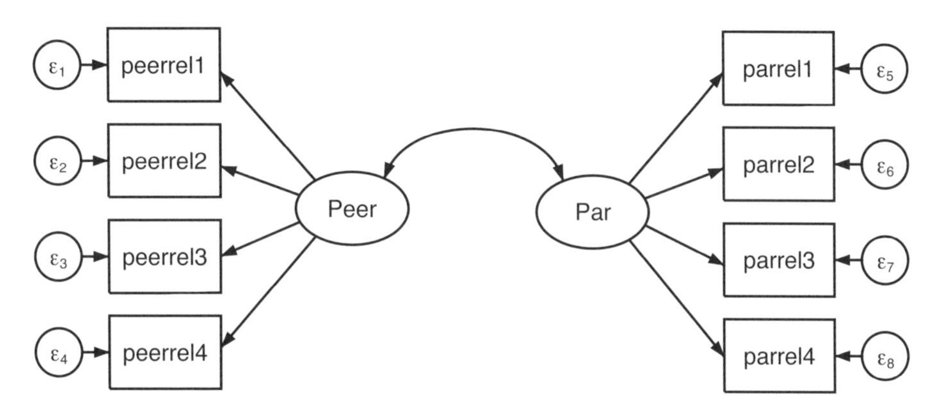

which, in command syntax, can be written

```
(Peer -> peerrel1 peerrel2 peerrel3 peerrel4)    ///
(Par  -> parrel1  parrel2  parrel3  parrel4)
```

We are using the same data used in [SEM] **example 15**, but we are using more of the data and fitting a different model. To remind you, those data were collected from students in grade 5. The dataset we are using, however, has data for students from grade 4 and from grade 5, which was created in [SEM] **example 19**. We have the following observed variables:

1. Four measures of physical ability.
2. Four measures of appearance.
3. Four measures of quality of relationship with peers.
4. Four measures of quality of relationship with parents.

In this example, we will consider solely the measurement problem, and include only the measurement variables (3) and (4). We are going to treat (3) as measures of underlying factor Peer and (4) as measures of underlying factor Par.

Below we will

1. Fit the model using all the data. This amounts to assuming that the students in grades 4 and 5 are identical in terms of this measurement problem.
2. Fit the model with sem's group() option, which will constrain some parameters to be the same for students in grades 4 and 5, and leave free of constraint the others.

Fitting the model using all the data

Throughout this example, we want you to appreciate that we are using summary statistics data and that matters not at all. Not one command would have a different syntax or option, or produce a different result, if we had the real data.

We begin by fitting the model using all the data:

```
. sem (Peer -> peerrel1 peerrel2 peerrel3 peerrel4)
>     (Par  -> parrel1  parrel2  parrel3  parrel4)
Endogenous variables

Measurement:  peerrel1 peerrel2 peerrel3 peerrel4 parrel1 parrel2 parrel3
              parrel4

Exogenous variables

Latent:       Peer Par

Fitting target model:

Iteration 0:   log likelihood =  -5559.545
Iteration 1:   log likelihood =  -5558.609
Iteration 2:   log likelihood = -5558.6017
Iteration 3:   log likelihood = -5558.6017

Structural equation model                       Number of obs      =       385
Estimation method  = ml
Log likelihood     = -5558.6017

 ( 1)  [peerrel1]Peer = 1
 ( 2)  [parrel1]Par = 1

------------------------------------------------------------------------------
             |                 OIM
             |      Coef.   Std. Err.      z    P>|z|     [95% Conf. Interval]
-------------+----------------------------------------------------------------
Measurement  |
  peerr~1 <- |
        Peer |          1  (constrained)
       _cons |   8.681221   .0937197    92.63   0.000     8.497534    8.864908
  -----------+----------------------------------------------------------------
  peerr~2 <- |
        Peer |   1.113865     .09796    11.37   0.000     .9218666    1.305863
       _cons |   7.828623   .1037547    75.45   0.000     7.625268    8.031979
  -----------+----------------------------------------------------------------
  peerr~3 <- |
        Peer |    1.42191    .114341    12.44   0.000     1.197806    1.646014
       _cons |   7.359896   .1149905    64.00   0.000     7.134519    7.585273
  -----------+----------------------------------------------------------------
  peerr~4 <- |
        Peer |   1.204146   .0983865    12.24   0.000     1.011312     1.39698
       _cons |   8.150779   .1023467    79.64   0.000     7.950183    8.351375
  -----------+----------------------------------------------------------------
  parrel1 <- |
         Par |          1  (constrained)
       _cons |   9.339558   .0648742   143.96   0.000     9.212407     9.46671
  -----------+----------------------------------------------------------------
  parrel2 <- |
         Par |   1.112383   .1378687     8.07   0.000     .8421655    1.382601
       _cons |   9.220494   .0742356   124.21   0.000     9.074994    9.365993
  -----------+----------------------------------------------------------------
  parrel3 <- |
         Par |   2.037924    .204617     9.96   0.000     1.636882    2.438966
       _cons |   8.676961    .088927    97.57   0.000     8.502667    8.851255
  -----------+----------------------------------------------------------------
```

```
    parrel4 <-
           Par      1.52253   .1536868     9.91   0.000    1.221309     1.82375
         _cons     9.045247   .0722358   125.22   0.000    8.903667    9.186826

Variance
    e.peerrel1     1.809309   .1596546                     1.521956    2.150916
    e.peerrel2     2.193804    .194494                     1.843884    2.610129
    e.peerrel3     1.911874    .214104                     1.535099    2.381126
    e.peerrel4     1.753037   .1749613                     1.441575    2.131792
     e.parrel1     1.120333   .0899209                     .9572541    1.311193
     e.parrel2     1.503003   .1200739                     1.285162    1.757769
     e.parrel3     .9680081   .1419777                     .7261617    1.290401
     e.parrel4     .8498834   .0933687                      .685245    1.054078
          Peer     1.572294   .2255704                     1.186904    2.082822
           Par     .5000022    .093189                     .3469983    .7204709

Covariance
    Peer
           Par     .4226706   .0725253     5.83   0.000    .2805236    .5648176

LR test of model vs. saturated: chi2(19)   =     28.19, Prob > chi2 = 0.0798
```

Notes:

1. We are using summary statistics data with data for two separate groups. There is no hint of that in the output above because sem combined the summary statistics and produced overall results just as if we had the real data.

Fitting the model using the group() option

```
. sem (Peer -> peerrel1 peerrel2 peerrel3 peerrel4)
>     (Par  -> parrel1  parrel2  parrel3  parrel4), group(grade)
Endogenous variables
Measurement:  peerrel1 peerrel2 peerrel3 peerrel4 parrel1 parrel2 parrel3
              parrel4
Exogenous variables
Latent:       Peer Par
Fitting target model:
Iteration 0:   log likelihood =  -13049.77  (not concave)
Iteration 1:   log likelihood = -11439.068  (not concave)
Iteration 2:   log likelihood =  -9825.231  (not concave)
Iteration 3:   log likelihood = -7019.5368  (not concave)
Iteration 4:   log likelihood = -6538.9872  (not concave)
Iteration 5:   log likelihood = -5920.2639  (not concave)
Iteration 6:   log likelihood = -5755.4178  (not concave)
Iteration 7:   log likelihood = -5663.0752  (not concave)
Iteration 8:   log likelihood = -5620.6701  (not concave)
Iteration 9:   log likelihood =  -5598.748  (not concave)
Iteration 10:  log likelihood = -5578.4839  (not concave)
Iteration 11:  log likelihood = -5572.6669
Iteration 12:  log likelihood = -5558.8811
Iteration 13:  log likelihood =  -5544.936
Iteration 14:  log likelihood = -5542.8249
Iteration 15:  log likelihood = -5542.6777
Iteration 16:  log likelihood = -5542.6774
Structural equation model                       Number of obs      =       385
Grouping variable  = grade                      Number of groups   =         2
Estimation method  = ml
Log likelihood     = -5542.6774
```

```
 ( 1)  [peerrel1]1bn.grade#c.Peer = 1
 ( 2)  [peerrel2]1bn.grade#c.Peer - [peerrel2]2.grade#c.Peer = 0
 ( 3)  [peerrel3]1bn.grade#c.Peer - [peerrel3]2.grade#c.Peer = 0
 ( 4)  [peerrel4]1bn.grade#c.Peer - [peerrel4]2.grade#c.Peer = 0
 ( 5)  [parrel1]1bn.grade#c.Par = 1
 ( 6)  [parrel2]1bn.grade#c.Par - [parrel2]2.grade#c.Par = 0
 ( 7)  [parrel3]1bn.grade#c.Par - [parrel3]2.grade#c.Par = 0
 ( 8)  [parrel4]1bn.grade#c.Par - [parrel4]2.grade#c.Par = 0
 ( 9)  [peerrel1]1bn.grade - [peerrel1]2.grade = 0
 (10)  [peerrel2]1bn.grade - [peerrel2]2.grade = 0
 (11)  [peerrel3]1bn.grade - [peerrel3]2.grade = 0
 (12)  [peerrel4]1bn.grade - [peerrel4]2.grade = 0
 (13)  [parrel1]1bn.grade - [parrel1]2.grade = 0
 (14)  [parrel2]1bn.grade - [parrel2]2.grade = 0
 (15)  [parrel3]1bn.grade - [parrel3]2.grade = 0
 (16)  [parrel4]1bn.grade - [parrel4]2.grade = 0
 (17)  [peerrel1]2.grade#c.Peer = 1
 (18)  [parrel1]2.grade#c.Par = 1
 (19)  [mean(Peer)]1bn.grade = 0
 (20)  [mean(Par)]1bn.grade = 0

------------------------------------------------------------------------------
             |                 OIM
             |      Coef.   Std. Err.      z    P>|z|     [95% Conf. Interval]
-------------+----------------------------------------------------------------
Measurement  |
  peerr~1 <- |
    Peer     |
         [*] |          1  (constrained)
    _cons    |
         [*] |   8.466537   .1473456    57.46   0.000     8.177745    8.755329
  -----------+----------------------------------------------------------------
  peerr~2 <- |
    Peer     |
         [*] |   1.109227   .0975265    11.37   0.000     .9180782    1.300375
    _cons    |
         [*] |   7.589871   .1632146    46.50   0.000     7.269976    7.909766
  -----------+----------------------------------------------------------------
  peerr~3 <- |
    Peer     |
         [*] |   1.409351   .1138295    12.38   0.000     1.186249    1.632453
    _cons    |
         [*] |   7.056996   .1964299    35.93   0.000        6.672    7.441991
  -----------+----------------------------------------------------------------
  peerr~4 <- |
    Peer     |
         [*] |   1.195974   .0980257    12.20   0.000     1.003847    1.388101
    _cons    |
         [*] |   7.893579    .169158    46.66   0.000     7.562036    8.225123
  -----------+----------------------------------------------------------------
  parrel1 <- |
    Par      |
         [*] |          1  (constrained)
    _cons    |
         [*] |   9.368654   .0819489   114.32   0.000     9.208037    9.529271
  -----------+----------------------------------------------------------------
  parrel2 <- |
    Par      |
         [*] |   1.104354   .1369358     8.06   0.000     .8359644    1.372743
    _cons    |
         [*] |   9.287629   .0903296   102.82   0.000     9.110587    9.464672
  -----------+----------------------------------------------------------------
```

parrel3 <-						
Par						
[*]	2.058586	.206057	9.99	0.000	1.654722	2.46245
_cons						
[*]	8.741898	.136612	63.99	0.000	8.474143	9.009653
parrel4 <-						
Par						
[*]	1.526703	.1552476	9.83	0.000	1.222424	1.830983
_cons						
[*]	9.096609	.1061607	85.69	0.000	8.888538	9.30468
Mean						
Peer						
1	0	(constrained)				
2	.3296867	.1570212	2.10	0.036	.0219307	.6374427
Par						
1	0	(constrained)				
2	-.0512438	.0818257	-0.63	0.531	-.2116191	.1091316
Variance						
e.peerrel1						
1	1.824185	.2739438			1.359068	2.44848
2	1.77381	.1889103			1.439641	2.185545
e.peerrel2						
1	2.236973	.3310871			1.673699	2.989815
2	2.165228	.2321563			1.75484	2.671589
e.peerrel3						
1	1.907009	.3383285			1.346909	2.70002
2	1.95068	.2586192			1.504301	2.529517
e.peerrel4						
1	1.639882	.2727635			1.183671	2.271925
2	1.822449	.2151823			1.445943	2.296991
e.parrel1						
1	.9669118	.1302489			.7425486	1.259067
2	1.213158	.1192634			1.000547	1.470949
e.parrel2						
1	.9683877	.133192			.7395628	1.268012
2	1.79031	.1747374			1.478595	2.167739
e.parrel3						
1	.8377564	.1986086			.526407	1.333257
2	1.015707	.1713758			.7297077	1.4138
e.parrel4						
1	.8343032	.1384648			.6026354	1.15503
2	.8599647	.1165864			.6592988	1.121706
Peer						
1	2.039327	.3784579			1.417493	2.93395
2	1.307988	.2061577			.9603778	1.781416
Par						
1	.4493011	.1011564			.2889991	.6985194
2	.5201706	.1029349			.3529427	.7666328
Covariance						
Peer						
Par						
1	.5012149	.1193345	4.20	0.000	.2673236	.7351063
2	.3867178	.0794552	4.87	0.000	.2309885	.5424471

Note: [*] identifies parameter estimates constrained to be equal across groups.

LR test of model vs. saturated: chi2(50) = 61.91, Prob > chi2 = 0.1204

Notes:

1. In *Which parameters vary by default, and which do not* in [SEM] **intro 5**, we wrote that, generally speaking, when we specify `group(`*groupvar*`)`, the measurement part of the model is constrained by default to be the same across the groups, whereas the remaining parts will have separate parameters for each group.

   More precisely, we revealed that `sem` classifies each parameter into one of nine classes, which are the following:

Class description	Class name	
1. structural coefficients	`scoef`	
2. structural intercepts	`scons`	
3. measurement coefficients	`mcoef`	
4. measurement intercepts	`mcons`	
5. covariances of structural errors	`serrvar`	
6. covariances of measurement errors	`merrvar`	
7. covariances between structural and measurement errors	`smerrcov`	
8. means of exogenous variables	`meanex`	(*)
9. covariances of exogenous variables	`covex`	(*)
10. all of the above	`all`	(*)
11. none of the above	`none`	

   (*) Be aware that classes 8, 9, and 10 (`meanex`, `covex`, and `all`) exclude the observed exogenous variables—include only the latent exogenous variables—unless you specify option `noxconditional` or the `noxconditional` option is otherwise implied; see [SEM] **sem option noxconditional**. This is what you would desire in most cases.

   By default, classes 3 and 4 are constrained to be equal and the rest are allowed to vary.

2. Thus you might expect that most of the parameters of our model would have been left unconstrained until you remember that we are fitting a measurement model. That is why `sem` listed 20 constraints at the top of the estimation results. Some of the constraints are substantive and some are normalization.

3. In the output, paths listed with an asterisk are constrained to be equal across groups.

   Paths labeled with group 1 and group 2 are group specific (unconstrained).

   In our data, group 1 corresponds with students in grade 4, and group 2 corresponds with students in grade 5.

4. It may surprise you that the output contains estimates for the means of the latent variables. Usually, `sem` does not report this.

   Usually, you are running on only one group of data and those means cannot be estimated, at least not without additional identifying constraints. When you are running on two or more groups, the means for all the groups except one can be estimated.

In [SEM] **example 21**, we use `estat ggof` to evaluate goodness of fit group by group.

In [SEM] **example 22**, we use `estat ginvariant` to test whether parameters that are constrained across groups should not be and whether parameters that are not constrained could be.

In [SEM] **example 23**, we show how to constrain the parameters we choose to be equal across groups.

Also see

[SEM] **example 3** — Two-factor measurement model

[SEM] **example 19** — Creating multiple-group summary statistics data

[SEM] **example 21** — Group-level goodness of fit

[SEM] **example 22** — Testing parameter equality across groups

[SEM] **example 23** — Specifying parameter constraints across groups

[SEM] **intro 5** — Comparing groups

[SEM] **sem** — Structural equation model estimation command

[SEM] **sem group options** — Fitting models on different groups

Title

example 21 — Group-level goodness of fit

Description

Below we demonstrate the `estat ggof` command, which may be used after `sem` with the `group()` option. `estat ggof` displays group-by-group goodness-of-fit statistics.

We pick up where [SEM] **example 20** left off:

```
. use http://www.stata-press.com/data/r12/sem_2fmmby
. sem (Peer -> peerrel1 peerrel2 peerrel3 peerrel4)    ///
      (Par  -> parrel1  parrel2  parrel3  parrel4), group(grade)
```

Remarks

```
. estat ggof
Group-level fit statistics
```

	N	SRMR	CD
grade			
1	134	0.088	0.969
2	251	0.056	0.955

```
Note: group-level chi-squared are not
  reported because of constraints between
  groups.
```

Notes:

1. Reported are the goodness-of-fit tests that `estat gof, stats(residuals)` would report. The difference is that they are reported for each group rather than overall.
2. If the fit is good, then SRMR (standardized root mean squared residual) will be close to 0 and CD (the coefficient of determination) will be near 1.

It is also appropriate to run `estat gof` to obtain overall results:

```
. estat gof, stats(residuals)
```

Fit statistic	Value	Description
Size of residuals		
SRMR	0.074	Standardized root mean squared residual
CD	0.958	Coefficient of determination

Also see

[SEM] **example 20** — Two-factor measurement model by group

[SEM] **example 4** — Goodness-of-fit statistics

[SEM] **estat ggof** — Group-level goodness-of-fit statistics

[SEM] **estat gof** — Goodness-of-fit statistics

Title

example 22 — Testing parameter equality across groups

Description

Below we demonstrate `estat ginvariant` to test parameters across groups.

We pick up where [SEM] **example 20** left off:

```
. use http://www.stata-press.com/data/r12/sem_2fmmby
. sem (Peer -> peerrel1 peerrel2 peerrel3 peerrel4)   ///
      (Par  -> parrel1  parrel2  parrel3  parrel4), group(grade)
```

Remarks

We use `estat ginvariant` to test whether parameters that are constrained to be equal across groups should not be and whether parameters that are not constrained across groups could be.

```
. estat ginvariant
```

Tests for group invariance of parameters

	Wald Test			Score Test		
	chi2	df	p>chi2	chi2	df	p>chi2
Measurement						
peerr~1 <-						
Peer	.	.	.	2.480	1	0.1153
_cons	.	.	.	0.098	1	0.7537
peerr~2 <-						
Peer	.	.	.	0.371	1	0.5424
_cons	.	.	.	0.104	1	0.7473
peerr~3 <-						
Peer	.	.	.	2.004	1	0.1568
_cons	.	.	.	0.002	1	0.9687
peerr~4 <-						
Peer	.	.	.	0.239	1	0.6246
_cons	.	.	.	0.002	1	0.9611
parrel1 <-						
Par	.	.	.	0.272	1	0.6019
_cons	.	.	.	0.615	1	0.4329
parrel2 <-						
Par	.	.	.	0.476	1	0.4903
_cons	.	.	.	3.277	1	0.0703
parrel3 <-						
Par	.	.	.	3.199	1	0.0737
_cons	.	.	.	1.446	1	0.2291
parrel4 <-						
Par	.	.	.	2.969	1	0.0849
_cons	.	.	.	0.397	1	0.5288

Mean						
Peer	.	.	.	.	.	.
Par	.	.	.	.	.	.
Variance						
e.peerrel1	0.024	1	0.8772	.	.	.
e.peerrel2	0.033	1	0.8565	.	.	.
e.peerrel3	0.011	1	0.9152	.	.	.
e.peerrel4	0.294	1	0.5879	.	.	.
e.parrel1	1.981	1	0.1593	.	.	.
e.parrel2	14.190	1	0.0002	.	.	.
e.parrel3	0.574	1	0.4486	.	.	.
e.parrel4	0.022	1	0.8813	.	.	.
Peer	4.583	1	0.0323	.	.	.
Par	0.609	1	0.4350	.	.	.
Covariance						
Peer						
Par	0.780	1	0.3772	.	.	.

Notes:

1. In the output above, score tests are reported for parameters that were constrained. The null hypothesis is that the constraint is valid. None of the tests reject a valid constraint.
2. Wald tests are reported for parameters that were not constrained. The null hypothesis is that a constraint would be valid. Only in two cases does it appear that grade 4 differs from grade 5, namely, the variance of `e.parrel2` and the variance of `Peer`.
3. We remind you that these tests are marginal tests. That is, each test is intended to be interpreted separately. These are not joint tests of simultaneous imposition or relaxation of constraints. If you want simultaneous tests, you must do them yourself using, for instance, the `test` command.

   If joint tests of parameter classes are desired, the `class` option can be used.

These results imply that none of the constraints we impose should be relaxed, and that perhaps we could constrain all the variances and covariances to be equal across groups except for the variances of `e.parrel2` and `Peer`. We do that in [SEM] **example 23**.

Also see

[SEM] **example 20** — Two-factor measurement model by group

[SEM] **example 23** — Specifying parameter constraints across groups

[SEM] **estat ginvariant** — Tests for invariance of parameters across groups

Title

example 23 — Specifying parameter constraints across groups

Description

Below we demonstrate how to constrain the parameters we want constrained to be equal across groups when using `sem` with the `group()` option.

We pick up where [SEM] **example 22** left off:

```
. use http://www.stata-press.com/data/r12/sem_2fmmby
. sem (Peer -> peerrel1 peerrel2 peerrel3 peerrel4)   ///
      (Par  -> parrel1  parrel2  parrel3  parrel4), group(grade)
. estat ginvariant
```

The `estat ginvariant` command implied that perhaps we could constrain all the variances and covariances to be equal across groups except for the variances of `e.parrel2` and `Peer`.

Remarks

Remarks are presented under the following headings:

Background
Fitting the constrained model

Background

We can specify which parameters we wish to allow to vary. Remember that `sem`'s `group()` option classifies the parameters of the model as

Class description	Class name	
1. structural coefficients	`scoef`	
2. structural intercepts	`scons`	
3. measurement coefficients	`mcoef`	
4. measurement intercepts	`mcons`	
5. covariances of structural errors	`serrvar`	
6. covariances of measurement errors	`merrvar`	
7. covariances between structural and measurement errors	`smerrcov`	
8. means of exogenous variables	`meanex`	(*)
9. covariances of exogenous variables	`covex`	(*)
10. all of the above	`all`	(*)
11. none of the above	`none`	

(*) Exogenous variables means just the latent exogenous variables unless you specify `sem` option `noxconditional` or you specify option `method(mlmv)` (which implies option `noxconditional`); see [SEM] **sem option noxconditional**.

When fitting a model with the `group()` option,

```
. sem ..., ... group(varname)
```

you may also specify the ginvariant() option:

```
. sem ..., ... group(varname) ginvariant(class names)
```

You may specify any of the class names as being ginvariant(). You may specify as many class names as you wish. When you specify ginvariant(), sem cancels its default actions on which parameters vary, and which do not, and which uses the information you specify. All classes that you do not mention as being ginvariant() are allowed to vary across groups.

By using ginvariant(), you can constrain, or free by your silence, whole classes of parameters. For instance, you could type

```
. sem ..., group(mygroup) ginvariant(mcoef mcons serrvar)
```

and you are constraining those parameters to be equal across groups and leaving unconstrained scoef, scons, merrvar, smerrcov, meanex, and covex.

In addition, if a class is constrained, you can still unconstrain individual coefficients. Consider the model

```
. sem ... (x1<-L) ...
```

If you typed

```
. sem ... (1: x1<-L@a1) (2: x1<-L@a2) ..., group(mygroup) ginvariant(all)
```

then all estimated parameters would be the same across groups except for the path x1<-L, and it would be free to vary in groups 1 and 2.

By the same token, if a class is unconstrained, you can still constrain individual coefficients. If you typed

```
. sem ... (1: x1<-L@a) (2: x1<-L@a) ..., group(mygroup) ginvariant(none)
```

then you would leave unconstrained all parameters except the path x1<-L, and it would be constrained to be equal in groups 1 and 2.

All this is discussed in [SEM] **intro 5**, including how to constrain and free variance and covariance parameters.

Fitting the constrained model

In our case, we wish to fit our model:

```
. sem (Peer -> peerrel1 peerrel2 peerrel3 peerrel4)    ///
      (Par  -> parrel1  parrel2  parrel3  parrel4),    ///
                group(grade)
```

We impose constraints on all parameters except the variances of `e.parrel2` and `Peer`. We can do that by typing

```
. sem (Peer -> peerrel1 peerrel2 peerrel3 peerrel4)
>     (Par  -> parrel1  parrel2  parrel3  parrel4),
>                 group(grade)
>                 ginvariant(all)
>                 var(1: e.parrel2@v1)
>                 var(2: e.parrel2@v2)
>                 var(1: Peer@v3)
>                 var(2: Peer@v4)
Endogenous variables

Measurement:  peerrel1 peerrel2 peerrel3 peerrel4 parrel1 parrel2 parrel3
              parrel4

Exogenous variables

Latent:       Peer Par

Fitting target model:

Iteration 0:   log likelihood = -5560.9934
Iteration 1:   log likelihood = -5552.6825
Iteration 2:   log likelihood = -5549.5679
Iteration 3:   log likelihood = -5549.3511
Iteration 4:   log likelihood = -5549.3501
Iteration 5:   log likelihood = -5549.3501

Structural equation model                       Number of obs      =       385
Grouping variable  = grade                      Number of groups   =         2
Estimation method  = ml
Log likelihood     = -5549.3501

 ( 1)  [peerrel1]1bn.grade#c.Peer = 1
 ( 2)  [peerrel2]1bn.grade#c.Peer - [peerrel2]2.grade#c.Peer = 0
 ( 3)  [peerrel3]1bn.grade#c.Peer - [peerrel3]2.grade#c.Peer = 0
 ( 4)  [peerrel4]1bn.grade#c.Peer - [peerrel4]2.grade#c.Peer = 0
 ( 5)  [parrel1]1bn.grade#c.Par = 1
 ( 6)  [parrel2]1bn.grade#c.Par - [parrel2]2.grade#c.Par = 0
 ( 7)  [parrel3]1bn.grade#c.Par - [parrel3]2.grade#c.Par = 0
 ( 8)  [parrel4]1bn.grade#c.Par - [parrel4]2.grade#c.Par = 0
 ( 9)  [var(e.peerrel1)]1bn.grade - [var(e.peerrel1)]2.grade = 0
 (10)  [var(e.peerrel2)]1bn.grade - [var(e.peerrel2)]2.grade = 0
 (11)  [var(e.peerrel3)]1bn.grade - [var(e.peerrel3)]2.grade = 0
 (12)  [var(e.peerrel4)]1bn.grade - [var(e.peerrel4)]2.grade = 0
 (13)  [var(e.parrel1)]1bn.grade - [var(e.parrel1)]2.grade = 0
 (14)  [var(e.parrel3)]1bn.grade - [var(e.parrel3)]2.grade = 0
 (15)  [var(e.parrel4)]1bn.grade - [var(e.parrel4)]2.grade = 0
 (16)  [cov(Peer,Par)]1bn.grade - [cov(Peer,Par)]2.grade = 0
 (17)  [var(Par)]1bn.grade - [var(Par)]2.grade = 0
 (18)  [peerrel1]1bn.grade - [peerrel1]2.grade = 0
 (19)  [peerrel2]1bn.grade - [peerrel2]2.grade = 0
 (20)  [peerrel3]1bn.grade - [peerrel3]2.grade = 0
 (21)  [peerrel4]1bn.grade - [peerrel4]2.grade = 0
 (22)  [parrel1]1bn.grade - [parrel1]2.grade = 0
 (23)  [parrel2]1bn.grade - [parrel2]2.grade = 0
 (24)  [parrel3]1bn.grade - [parrel3]2.grade = 0
 (25)  [parrel4]1bn.grade - [parrel4]2.grade = 0
 (26)  [peerrel1]2.grade#c.Peer = 1
 (27)  [parrel1]2.grade#c.Par = 1
```

	Coef.	OIM Std. Err.	z	P>|z|	[95% Conf.	Interval]
Measurement						
peerr~1 <-						
Peer						
[*]	1	(constrained)				
_cons						
[*]	8.708274	.0935844	93.05	0.000	8.524852	8.891696
peerr~2 <-						
Peer						
[*]	1.112225	.0973506	11.43	0.000	.9214217	1.303029
_cons						
[*]	7.858713	.1035989	75.86	0.000	7.655663	8.061763
peerr~3 <-						
Peer						
[*]	1.416486	.113489	12.48	0.000	1.194052	1.638921
_cons						
[*]	7.398217	.1147474	64.47	0.000	7.173316	7.623118
peerr~4 <-						
Peer						
[*]	1.196494	.0976052	12.26	0.000	1.005191	1.387796
_cons						
[*]	8.183148	.1021513	80.11	0.000	7.982936	8.383361
parrel1 <-						
Par						
[*]	1	(constrained)				
_cons						
[*]	9.339558	.0648742	143.96	0.000	9.212407	9.46671
parrel2 <-						
Par						
[*]	1.100315	.1362999	8.07	0.000	.8331722	1.367458
_cons						
[*]	9.255299	.0725417	127.59	0.000	9.11312	9.397478
parrel3 <-						
Par						
[*]	2.051278	.2066714	9.93	0.000	1.64621	2.456347
_cons						
[*]	8.676961	.088927	97.57	0.000	8.502667	8.851255
parrel4 <-						
Par						
[*]	1.529938	.154971	9.87	0.000	1.2262	1.833675
_cons						
[*]	9.045247	.0722358	125.22	0.000	8.903667	9.186826

```
Variance
  e.peerrel1
         [*]    1.799133    .159059                        1.512898    2.139523
  e.peerrel2
         [*]    2.186953    .193911                        1.838086    2.602035
  e.peerrel3
         [*]    1.915661   .2129913                         1.54056    2.382094
  e.peerrel4
         [*]    1.767354   .1746104                         1.45622    2.144965
  e.parrel1
         [*]    1.125082   .0901338                        .9615942    1.316366
  e.parrel2
           1    .9603043     .13383                         .730775    1.261927
           2    1.799668   .1747351                        1.487807    2.176898
  e.parrel3
         [*]    .9606889   .1420406                        .7190021    1.283617
  e.parrel4
         [*]    .8496935   .0933448                        .6850966    1.053835
  Peer
           1    1.951555   .3387796                        1.388727    2.742489
           2    1.361431   .2122853                        1.002927    1.848084
  Par
         [*]    .4952527   .0927994                        .3430288    .7150281

Covariance
  Peer
    Par
         [*]    .4096197   .0708726     5.78   0.000       .2707118    .5485275

Note: [*] identifies parameter estimates constrained to be equal across
      groups.
LR test of model vs. saturated: chi2(61)  =     75.25, Prob > chi2 = 0.1037
```

Notes:

1. In [SEM] **example 20**, we previously fit this model by typing

```
. sem (...) (...), group(grade)
```

   This time, we typed

```
. sem (...) (...), group(grade)          ///
               ginvariant(all)           ///
               var(1: e.parrel2@v1)      ///
               var(2: e.parrel2@v2)      ///
               var(1: Peer@v3)           ///
               var(2: Peer@v4)
```

2. Previously, `sem, group()` mentioned 20 constraints that it imposed because of normalization or because of assumed `ginvariant(mcoef mcons)`.

   This time, `sem, group()` mentioned 27 constraints. It applied more constraints because we specified `ginvariant(all)`.

3. After the `ginvariant(all)` option, we relaxed the following constraints:

```
var(1: e.parrel2@v1)
var(2: e.parrel2@v2)
var(1: Peer@v3)
var(2: Peer@v4)
```

   `ginvariant(all)` specified, among other constraints, that

```
var(1: e.parrel2) == var(2: e.parrel2)
var(1: Peer) == var(2: Peer)
```

ginvariant(all) did that by secretly issuing the options

```
var(1: e.parrel2@secretname1)
var(2: e.parrel2@secretname1)
var(1: Peer@secretname2)
var(2: Peer@secretname2)
```

because that is how you impose equality constraints using the path notation. When we specified

```
var(1: e.parrel2@v1)
var(2: e.parrel2@v2)
var(1: Peer@v3)
var(2: Peer@v4)
```

our new constraints overrode the secretly issued constraints. It would not have worked to leave off the symbolic names; see *Added syntax when option group() is specified* in [SEM] **sem path notation**. We specified the symbolic names v1, v2, v3, and v4. v1 and v2 overrode *secretname1*, and thus the constraint that var(e.parrel2) be equal across the two groups was relaxed. v3 and v4 overrode *secretname2*, and the constraint that var(Peer) be equal across groups was relaxed.

Also see

[SEM] **example 20** — Two-factor measurement model by group

[SEM] **example 22** — Testing parameter equality across groups

[SEM] **intro 5** — Comparing groups

[SEM] **sem group options** — Fitting models on different groups

Title

example 24 — Reliability

Description

Below we demonstrate sem's reliability() option using the following data:

```
. use http://www.stata-press.com/data/r12/sem_rel
(measurement error with known reliabilities)
. summarize
```

Variable	Obs	Mean	Std. Dev.	Min	Max
y	1234	701.081	71.79378	487	943
x1	1234	100.278	14.1552	51	149
x2	1234	100.2066	14.50912	55	150

```
. notes
_dta:
  1.  Fictional data.
  2.  Variables x1 and x2 each contain a test score designed to measure X.  The
      test is scored to have mean 100.
  3.  Variables x1 and x2 are both known to have reliability 0.5.
  4.  Variable y is the outcome, believed to be related to X.
```

See [SEM] **sem option reliability()** for background.

Remarks

Remarks are presented under the following headings:

Baseline model (reliability ignored)

```
. sem (y <- x1)
Endogenous variables
Observed:  y
Exogenous variables
Observed:  x1
Fitting target model:
Iteration 0:   log likelihood = -11629.745
Iteration 1:   log likelihood = -11629.745
Structural equation model                        Number of obs      =       1234
Estimation method  = ml
Log likelihood     = -11629.745
```

	Coef.	OIM Std. Err.	z	P>|z|	[95% Conf.	Interval]
Structural						
y <-						
x1	3.54976	.1031254	34.42	0.000	3.347637	3.751882
_cons	345.1184	10.44365	33.05	0.000	324.6492	365.5876
Variance						
e.y	2627.401	105.7752			2428.053	2843.115

```
LR test of model vs. saturated: chi2(0)   =      0.00, Prob > chi2 =      .
```

Notes:

1. In these data, variable x1 is measured with error.
2. If we ignore that, we obtain a path coefficient for y<-x1 of 3.55.
3. We also ran this model for y<-x2. We obtained a path coefficient of 3.48.

Model with reliability

```
. sem (x1<-X) (y<-X), reliability(x1 .5)
Endogenous variables
Measurement:  x1 y
Exogenous variables
Latent:       X
Fitting target model:
Iteration 0:   log likelihood = -11745.845
Iteration 1:   log likelihood = -11661.626
Iteration 2:   log likelihood = -11631.469
Iteration 3:   log likelihood = -11629.755
Iteration 4:   log likelihood = -11629.745
Iteration 5:   log likelihood = -11629.745
Structural equation model                       Number of obs      =      1234
Estimation method  = ml
Log likelihood     = -11629.745
 ( 1)  [x1]X = 1
 ( 2)  [var(e.x1)]_cons = 100.1036

------------------------------------------------------------------------------
             |                 OIM
             |      Coef.   Std. Err.      z    P>|z|     [95% Conf. Interval]
-------------+----------------------------------------------------------------
Measurement  |
  x1 <-      |
           X |          1  (constrained)
       _cons |    100.278   .4027933   248.96   0.000      99.4885    101.0674
  -----------+----------------------------------------------------------------
  y <-       |
           X |    7.09952    .352463    20.14   0.000     6.408705    7.790335
       _cons |    701.081   2.042929   343.17   0.000      697.077    705.0851
-------------+----------------------------------------------------------------
Variance     |
        e.x1 |   100.1036  (constrained)
         e.y |    104.631   207.3381                      2.152334    5086.411
           X |   100.1036   8.060038                      85.48963    117.2157
------------------------------------------------------------------------------
LR test of model vs. saturated: chi2(0)   =      0.00, Prob > chi2 =      .
```

Notes:

1. We wish to estimate the effect of `y<-x1` when `x1` is measured with error (0.50 reliability). To do that, we introduce latent variable `X` and write our model as `(x1<-X) (y<-X)`.

2. When we ignored the measurement error of `x1`, we obtained a path coefficient for `y<-x1` of 3.55. Taking into account the measurement error, we obtain a coefficient of 7.1.

Model with two measurement variables and reliability

```
. sem (x1 x2<-X) (y<-X), reliability(x1 .5  x2 .5)
Endogenous variables
Measurement:  x1 x2 y
Exogenous variables
Latent:       X
Fitting target model:
Iteration 0:   log likelihood = -16258.636
Iteration 1:   log likelihood = -16258.401
Iteration 2:   log likelihood =   -16258.4
Structural equation model                        Number of obs      =      1234
Estimation method  = ml
Log likelihood     =   -16258.4
 ( 1)  [x1]X = 1
 ( 2)  [var(e.x1)]_cons = 100.1036
 ( 3)  [var(e.x2)]_cons = 105.1719
```

	Coef.	OIM Std. Err.	z	P>|z|	[95% Conf.	Interval]
Measurement						
x1 <-						
X	1	(constrained)				
_cons	100.278	.4037851	248.34	0.000	99.48655	101.0694
x2 <-						
X	1.030101	.0417346	24.68	0.000	.9483029	1.1119
_cons	100.2066	.4149165	241.51	0.000	99.39342	101.0199
y <-						
X	7.031299	.2484176	28.30	0.000	6.544409	7.518188
_cons	701.081	2.042928	343.17	0.000	697.077	705.0851
Variance						
e.x1	100.1036	(constrained)				
e.x2	105.1719	(constrained)				
e.y	152.329	105.26			39.31868	590.1553
X	101.0907	7.343656			87.67509	116.5591

```
LR test of model vs. saturated: chi2(2)   =      0.59, Prob > chi2 = 0.7430
```

Notes:

1. We wish to estimate the effect of `y<-X`. We have two measures of `X`, `x1` and `x2`, both measured with error (0.50 reliability).
2. In the previous section, we used just `x1`. We obtained path coefficient 7.1 with standard error 0.4.
3. Using both `x1` and `x2`, we obtain path coefficient 7.0 and standard error 0.2.
4. We at StataCorp created these fictional data. The true coefficient is 7.

Also see

[SEM] **sem option reliability()** — Fraction of variance not due to measurement error

[SEM] **example 1** — Single-factor measurement model

Title

example 25 — Creating summary statistics data from raw data

Description

Below we show how to create summary statistics data from raw data. We will use `auto.dta`, which surely needs no introduction:

```
. sysuse auto
(1978 Automobile Data)
. describe
  (output omitted )
. summarize
  (output omitted )
```

Remarks

Remarks are presented under the following headings:

Preparing data for conversion
Converting to summary statistics form
Publishing summary statistics data
Creating summary statistics data with multiple groups

We are going to create summary statistics data containing the variables `price`, `mpg`, `weight`, `displacement`, and `foreign`.

Preparing data for conversion

Before building the summary statistics data, prepare the data to be converted:

1. Drop variables that you do not intend to include in the summary statistics data. Dropping variables is not a requirement, but it will be easier to spot problems if you begin by eliminating the irrelevant variables.
2. Verify that you have no string variables in the resulting data. Summary statistics datasets cannot contain string values.
3. Verify that there are no missing values. If there are, be aware that observations containing one or more variables with missing values will be omitted from the summary statistics data.
4. Verify that all variables are on a reasonable scale. We recommend that the means of variables be only 3 or 4 orders of magnitude different from each other. This will help to preserve numerical accuracy when the summary statistics data are used.
5. Create any new variables containing transformations of existing variables that might be useful later. Once the data are converted to summary statistics form, you will not be able to create such variables.
6. Place the variables in a logical order. That will help the user of the summary statistics data understand the data.
7. Save the resulting prepared data. Probably you will never need the prepared data, but one never knows for sure.

We take our own advice below:

```
. * -------------------------------------------------------------------------
. * Suggestion 1:  Keep relevant variables:
. *
. keep price mpg weight displacement foreign

.
. * -------------------------------------------------------------------------
. * Suggestion 2: Check for string variables
. * Suggestion 3: Verify no missing values
. * Suggestion 4: Verify variables on a reasonable scale:
. *
. summarize

    Variable |       Obs        Mean    Std. Dev.       Min        Max
-------------+--------------------------------------------------------
       price |        74    6165.257    2949.496       3291      15906
         mpg |        74     21.2973    5.785503         12         41
      weight |        74    3019.459    777.1936       1760       4840
displacement |        74    197.2973    91.83722         79        425
     foreign |        74    .2972973    .4601885          0          1

.
. * We will rescale weight and price:
. replace    weight = weight/1000
weight was int now float
(74 real changes made)

. replace    price  =  price/1000
price was int now float
(74 real changes made)

. label var weight "Weight (1000s lbs.)"

. label var price  "Price  ($1,000s)"

. * and now we check our work:
. *
. summarize

    Variable |       Obs        Mean    Std. Dev.       Min        Max
-------------+--------------------------------------------------------
       price |        74    6.165257    2.949496      3.291     15.906
         mpg |        74     21.2973    5.785503         12         41
      weight |        74    3.019459    .7771936       1.76       4.84
displacement |        74    197.2973    91.83722         79        425
     foreign |        74    .2972973    .4601885          0          1

.
. * -------------------------------------------------------------------------
. * Suggestion 5: Create useful transformations:
. *
. gen       gpm    = 1/mpg

. label var gpm    "Gallons per mile"

.
```

```
. * ---------------------------------------------------------------------------
. * Suggestion 6: Place variables in logical order:
. *
. order price mpg gpm
. summarize
    Variable |       Obs        Mean    Std. Dev.       Min        Max
-------------+--------------------------------------------------------
       price |        74    6.165257    2.949496      3.291     15.906
         mpg |        74     21.2973    5.785503         12         41
         gpm |        74    .0501928    .0127986   .0243902   .0833333
      weight |        74    3.019459    .7771936       1.76       4.84
displacement |        74    197.2973    91.83722         79        425
-------------+--------------------------------------------------------
     foreign |        74    .2972973    .4601885          0          1
.
. * ---------------------------------------------------------------------------
. * Suggestion 7: save prepared data
. *
. save auto_raw
file auto_raw.dta saved
. * ---------------------------------------------------------------------------
```

Converting to summary statistics form

To create the summary statistics dataset, you just need to type `ssd build` followed by the names of the variables to be included. If you have previously kept the relevant variables, you can type `ssd build _all`.

We recommend the following steps:

1. Convert data to summary statistics form:

```
. ssd build _all
```

2. Review the result:

```
. ssd describe
. notes
. ssd list
```

3. Digitally sign thc data:

```
. datasignature set
```

4. Save the data:

```
. save auto_ss
```

We follow our advice below. After that, we will show you the advantages of digitally signing the data.

```
. * ---------------------------------------------------------------------------
. * Convert data:
. *
. ssd build _all
 (data in memory now summary statistics data; you can use ssc describe and
  ssd list to describe and list results.)
```

```
. * ---------------------------------------------------------------------------
. * Review results:
. *
. ssd describe

  Summary statistics data
    obs:              74
   vars:               6
                                          (_dta has notes)
  ----------------------------------------------------------------------------
  variable name                 variable label
  ----------------------------------------------------------------------------
  price                         Price  ($1,000s)
  mpg                           Mileage (mpg)
  gpm                           Gallons per mile
  weight                        Weight (1000s lbs.)
  displacement                  Displacement (cu. in.)
  foreign                       Car type
  ----------------------------------------------------------------------------

. notes

_dta:
  1.  summary statistics data built from 'auto_raw.dta' on 30 Jun 2011 15:32:33
      using -ssd build _all-

. ssd list

  Observations = 74

  Means:
          price          mpg          gpm       weight  displacement
      6.1652567    21.297297     .0501928    3.0194595      197.2973

        foreign
       .2972973

  Variances implicitly defined; they are the diagonal of the covariance
  matrix.

  Covariances:
          price          mpg          gpm       weight  displacement
      8.6995258
     -7.9962828    33.472047
      .02178417   -.06991586     .0001638
      1.2346748   -3.6294262    .00849897    .60402985
      134.06705   -374.92521    .90648519     63.87345     8434.0748
      .06612809    1.0473899   -.00212897   -.21202888    -25.938912

        foreign
      .21177342

. * ---------------------------------------------------------------------------
. * Digitally sign:
. *
. datasignature set
  8:8(102846):1810957193:2605039657       (data signature set)

. * ---------------------------------------------------------------------------
. * Save:
. *
. save auto_ss
file auto_ss.dta saved

. * ---------------------------------------------------------------------------
```

We recommend digitally signing the data. This way, anyone can verify later that the data are unchanged:

```
. datasignature confirm
  (data unchanged since 30jun2011 15:32)
```

Let us show you what would happen if the data had changed:

```
. replace mpg = mpg+.0001 in 5
(1 real change made)
. datasignature confirm
  data have changed since 30jun2011 15:34
r(9);
```

There is no reason for you or anyone else to change the summary statistics data after it has been created, so we recommend that you digitally sign the data. With regular datasets, users do make changes, if only by adding variables.

Be aware that the data signature is a function of the variable names, so if you rename a variable—something you are allowed to do—the signature will change and `datasignature` will report, for example, "data have changed since 30jun2011 15:34". Solutions to that problem are discussed in [SEM] **ssd**.

Publishing summary statistics data

The summary statistics dataset you have just created can obviously be sent to and used by any Stata user. If you wish to publish your data in printed form, use `ssd describe` and `ssd list` to describe and list the data.

Creating summary statistics data with multiple groups

The process for creating summary statistics data containing multiple groups is nearly the same as for creating single-group data. The only differences are (1) you do not drop the group variable during preparation, and (2) rather than typing

```
. ssd build _all
```

you type

```
. ssd build _all, group(varname)
```

Below we build the automobile summary statistics data again, but this time, we specify `group(rep78)`:

```
. ssd build _all, group(rep78)
```

If you think carefully about this, you may be worried that `_all` includes `rep78` and thus we will be including the grouping variable among the summary statistics. `ssd build` knows to omit the group variable:

```
. * -------------------------------------------------------------------------
. * Suggestion 1:  Keep relevant variables:
. *
. webuse auto
(1978 Automobile Data)
. keep price mpg weight displacement foreign rep78
.
```

```
. * ------------------------------------------------------------------------
. * Suggestion 2: Check for string variables
. * Suggestion 3: Verify no missing values
. * Suggestion 4: Verify variables on a reasonable scale:
. *
. summarize

    Variable |       Obs        Mean    Std. Dev.       Min        Max
-------------+--------------------------------------------------------
       price |        74    6165.257    2949.496       3291      15906
         mpg |        74     21.2973    5.785503         12         41
       rep78 |        69    3.405797    .9899323          1          5
      weight |        74    3019.459    777.1936       1760       4840
displacement |        74    197.2973    91.83722         79        425
-------------+--------------------------------------------------------
     foreign |        74    .2972973    .4601885          0          1
. drop if rep78 >= .
(5 observations deleted)
. * We will rescale weight and price:
. replace   weight = weight/1000
weight was int now float
(69 real changes made)
. replace   price  =  price/1000
price was int now float
(69 real changes made)
. label var weight "Weight (1000s lbs.)"
. label var price  "Price  ($1,000s)"
. * and now we check our work:
. summarize

    Variable |       Obs        Mean    Std. Dev.       Min        Max
-------------+--------------------------------------------------------
       price |        69    6.146043     2.91244      3.291     15.906
         mpg |        69    21.28986    5.866408         12         41
       rep78 |        69    3.405797    .9899323          1          5
      weight |        69    3.032029    .7928515       1.76       4.84
displacement |        69         198    93.14789         79        425
-------------+--------------------------------------------------------
     foreign |        69    .3043478    .4635016          0          1
.
. * ------------------------------------------------------------------------
. * Suggestion 5: Create useful transformations:
. *
. gen        gpm     = 1/mpg
. label var gpm     "Gallons per mile"
.
. * ------------------------------------------------------------------------
. * Suggestion 6: Place variables in logical order:
. *
. order price mpg gpm
. summarize

    Variable |       Obs        Mean    Std. Dev.       Min        Max
-------------+--------------------------------------------------------
       price |        69    6.146043     2.91244      3.291     15.906
         mpg |        69    21.28986    5.866408         12         41
         gpm |        69    .0502584    .0128353   .0243902   .0833333
       rep78 |        69    3.405797    .9899323          1          5
      weight |        69    3.032029    .7928515       1.76       4.84
-------------+--------------------------------------------------------
displacement |        69         198    93.14789         79        425
     foreign |        69    .3043478    .4635016          0          1
```

```
. * -------------------------------------------------------------------------
. * Suggestion 7: save prepared data
. *
. save auto_group_raw
file auto_group_raw.dta saved
. * -------------------------------------------------------------------------
.
. * -------------------------------------------------------------------------
. * Convert data:
. *
. ssd build _all, group(rep78)
 (data in memory now summary statistics data; you can use ssc describe and
  ssd list to describe and list results.)
. * -------------------------------------------------------------------------
. * Review results:
. *
. ssd describe
  Summary statistics data
    obs:                 69
   vars:                  6
                                          (_dta has notes)
  -----------------------------------------------------------------------
  variable name                  variable label
  -----------------------------------------------------------------------
  price                          Price  ($1,000s)
  mpg                            Mileage (mpg)
  gpm                            Gallons per mile
  weight                         Weight (1000s lbs.)
  displacement                   Displacement (cu. in.)
  foreign                        Car type
  -----------------------------------------------------------------------
  Group variable:  rep78  (5 groups)
   Obs. by group:  2, 8, 30, 18, 11
. notes
_dta:
  1.  summary statistics data built from 'auto_group_raw.dta' on 30 Jun 2011
      15:32:33 using -ssd build _all, group(rep78)-
. ssd list
-----------------------------------------------------------------------------
Group rep78==1:
  (output omitted)
-----------------------------------------------------------------------------
Group rep78==2:
  (output omitted)
-----------------------------------------------------------------------------
Group rep78==3:
  (output omitted)
-----------------------------------------------------------------------------
Group rep78==4:
  (output omitted)
-----------------------------------------------------------------------------
Group rep78==5:
  Observations = 11
  Means:
         price           mpg           gpm        weight  displacement
         5.913     27.363636     .04048131     2.3227273     111.09091

       foreign
     .81818182
```

```
  Variances implicitly defined; they are the diagonal of the covariance
  matrix.

  Covariances:
          price          mpg           gpm        weight  displacement
      6.8422143
     -15.608899    76.254545
      .02750797    -.1184875     .00019114
        .956802   -3.0610912     .00510833     .16856184
      55.493298   -201.03636     .33150758     9.9577283     648.09091
      .34169998   -.92727273     .00182175     .07254547     3.8181818

        foreign
      .16363636

. * ---------------------------------------------------------------------------
. * Digitally sign:
. *
. datasignature set
  40:8(34334):2463066775:1212422464        (data signature set)
. * ---------------------------------------------------------------------------
. * Save:
. *
. save auto_group_ss
file auto_group_ss.dta saved
. * ---------------------------------------------------------------------------
```

Also see

[SEM] **ssd** — Making summary statistics data

Title

example 26 — Fitting a model using data missing at random

Description

sem method(mlmv) is demonstrated using

```
. use http://www.stata-press.com/data/r12/cfa_missing
(CFA MAR data)

. summarize

    Variable |       Obs        Mean    Std. Dev.       Min        Max
-------------+--------------------------------------------------------
          id |       500       250.5    144.4818          1        500
       test1 |       406    97.37475    13.91442    56.0406   136.5672
       test2 |       413    98.04501    13.84145   62.25496   129.3881
       test3 |       443    100.9699     13.4862   65.51753   137.3046
       test4 |       417    99.56815    14.25438    53.8719   153.9779
-------------+--------------------------------------------------------
       taken |       500       3.358    .6593219          2          4

. notes

_dta:
  1.  Fictional data on 500 subjects taking four tests.
  2.  Tests results M.A.R. (missing at random).
  3.  230 took all 4 tests
  4.  219 took 3 of the 4 tests
  5.  51 took 2 of the 4 tests
  6.  All tests have expected mean 100, s.d. 14.
```

See *Assumptions and choice of estimation method* in [SEM] **intro 3** for background.

Remarks

Remarks are presented under the following headings:

Fitting the model using method(ml)
Fitting the model using method(mlmv)

Fitting the model using method(ml)

We fit a single-factor measurement model.

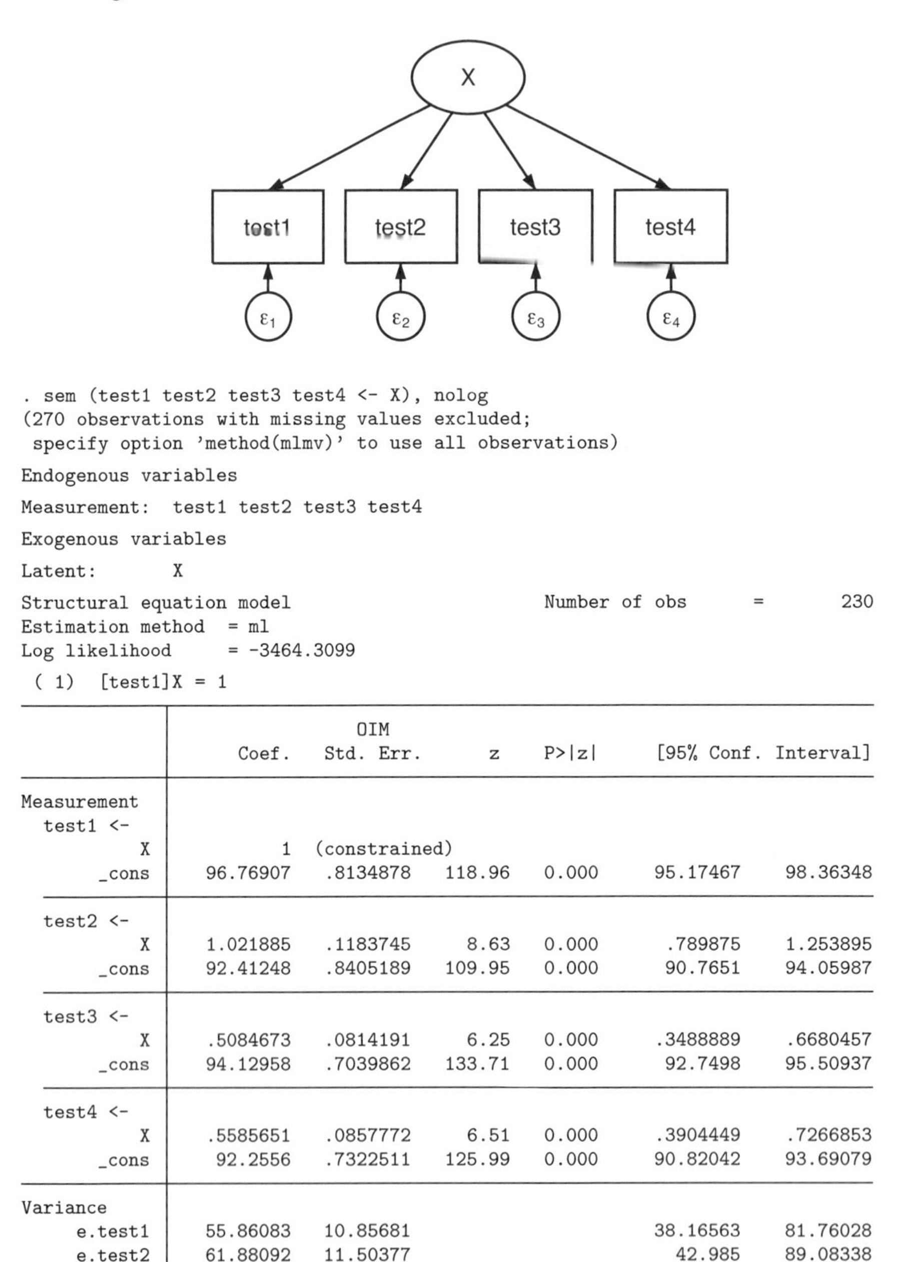


```
. sem (test1 test2 test3 test4 <- X), nolog
(270 observations with missing values excluded;
 specify option 'method(mlmv)' to use all observations)
Endogenous variables
Measurement:  test1 test2 test3 test4
Exogenous variables
Latent:       X
Structural equation model                       Number of obs      =       230
Estimation method  = ml
Log likelihood     = -3464.3099
 ( 1)  [test1]X = 1
------------------------------------------------------------------------------
             |                 OIM
             |      Coef.   Std. Err.      z    P>|z|     [95% Conf. Interval]
-------------+----------------------------------------------------------------
Measurement  |
  test1 <-   |
           X |          1  (constrained)
       _cons |   96.76907   .8134878   118.96   0.000     95.17467    98.36348
  -----------+----------------------------------------------------------------
  test2 <-   |
           X |   1.021885   .1183745     8.63   0.000      .789875    1.253895
       _cons |   92.41248   .8405189   109.95   0.000      90.7651    94.05987
  -----------+----------------------------------------------------------------
  test3 <-   |
           X |   .5084673   .0814191     6.25   0.000     .3488889    .6680457
       _cons |   94.12958   .7039862   133.71   0.000      92.7498    95.50937
  -----------+----------------------------------------------------------------
  test4 <-   |
           X |   .5585651   .0857772     6.51   0.000     .3904449    .7266853
       _cons |    92.2556   .7322511   125.99   0.000     90.82042    93.69079
-------------+----------------------------------------------------------------
Variance     |
     e.test1 |   55.86083   10.85681                      38.16563    81.76028
     e.test2 |   61.88092   11.50377                        42.985    89.08338
     e.test3 |   89.07839   8.962574                      73.13566    108.4965
     e.test4 |   93.26508   9.504276                      76.37945    113.8837
           X |   96.34453   16.28034                      69.18161    134.1725
------------------------------------------------------------------------------
LR test of model vs. saturated: chi2(2)   =      0.39, Prob > chi2 = 0.8212
```

Notes:

1. This model was fit using 230 of the 500 observations in the dataset. Unless you use `sem`'s `method(mlmv)`, observations are casewise omitted, meaning that if there is a single variable with a missing value among the variables being used, the observation is ignored.
2. The coefficients for `test3` and `test4` are 0.51 and 0.56. Because we at StataCorp manufactured these data, we can tell you that the true coefficients are 1.
3. The error variance for `e.test1` and `e.test2` are understated. These data were manufactured with an error variance of 100.
4. These data are missing at random (MAR), not missing completely at random (MCAR). In MAR data, which values are missing can be a function of the observed values in the data. MAR data can produce biased estimates if the missingness is ignored, as we just did. MCAR data do not bias estimates.

Fitting the model using method(mlmv)

```
. sem (test1 test2 test3 test4 <- X), method(mlmv) nolog
Endogenous variables

Measurement:  test1 test2 test3 test4

Exogenous variables

Latent:       X

Structural equation model                       Number of obs      =       500
Estimation method  = mlmv
Log likelihood     = -6592.9961

 ( 1)  [test1]X = 1
```

	Coef.	OIM Std. Err.	z	P>|z|	[95% Conf.	Interval]
Measurement						
test1 <-						
X	1	(constrained)				
_cons	98.94386	.6814418	145.20	0.000	97.60826	100.2795
test2 <-						
X	1.069952	.1079173	9.91	0.000	.8584378	1.281466
_cons	99.84218	.6911295	144.46	0.000	98.48759	101.1968
test3 <-						
X	.9489025	.0896098	10.59	0.000	.7732706	1.124534
_cons	101.0655	.6256275	161.54	0.000	99.83928	102.2917
test4 <-						
X	1.021626	.0958982	10.65	0.000	.8336687	1.209583
_cons	99.64509	.6730054	148.06	0.000	98.32603	100.9642
Variance						
e.test1	101.1135	10.1898			82.99057	123.1941
e.test2	95.45572	10.79485			76.47892	119.1413
e.test3	95.14847	9.053014			78.9611	114.6543
e.test4	101.0943	10.0969			83.12124	122.9536
X	94.04629	13.96734			70.29509	125.8225

```
LR test of model vs. saturated: chi2(2)    =        2.27, Prob > chi2 = 0.3209
```

Notes:

1. The model is now fit using all 500 observations in the dataset.
2. The coefficients for `test3` and `test4`—previously 0.51 and 0.56—are now 0.95 and 1.02.
3. Error variance estimates are now consistent with the true value of 100.
4. Standard errors of path coefficients are mostly smaller than reported in the previous model.
5. `method(mlmv)` requires that the data be MCAR or MAR.
6. `method(mlmv)` requires that the data be multivariate normal.

Also see

[SEM] **intro 3** — Substantive concepts

[SEM] **sem option method()** — Specifying method and calculation of VCE

Title

GUI — Graphical user interface

Description

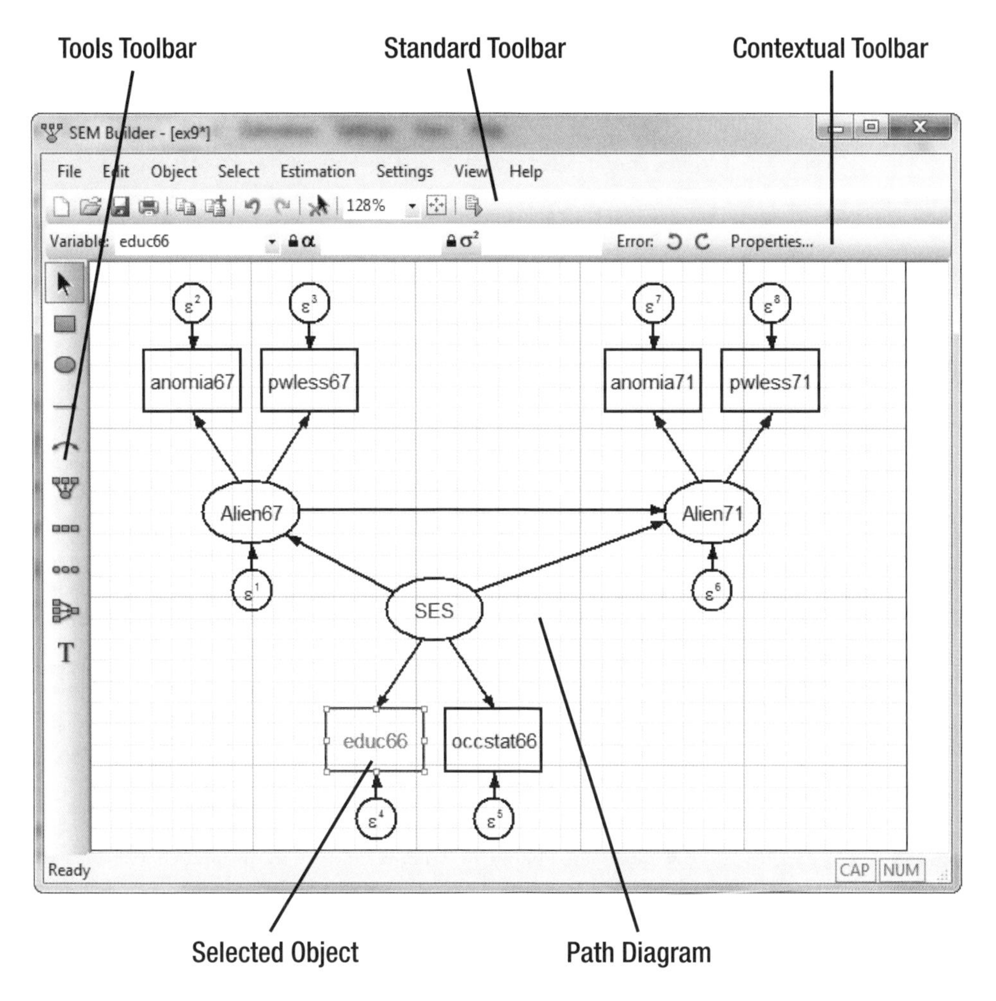

Launch the SEM Builder by selecting the menu item **Statistics > Structural equation modeling (SEM) > Model building and estimation**. You can also type `sembuilder` in the Command window.

The SEM Builder lets you build path diagrams for SEM models, fit those models from within the Builder, show results on the path diagram, and analyze the fitted model (modification indices, goodness of fit, direct and indirect effects, etc.).

Select the Add Observed Variable Tool, , and click within the diagram to add an observed variable. If the variable is not placed exactly where you want it, simply drag it to your preferred location. After adding the variable, use the **Variable** control in the Contextual Toolbar to select a variable from your dataset or type a variable name into the edit field.

Add latent variables to the model by using the tool. Type a name in the **Name** control of the Contextual Toolbar to name the latent variable.

Clicking within the diagram with either the or tool places a variable of the default size and shape. Hold the button and drag to place a variable of custom size and shape. Select the **Properties** button in the Contextual Toolbar or double-click on a variable to launch a dialog box where you can change even more properties of the variable. You can customize the size, shape, label font, and other appearance characteristics of the variable from the **Appearance** tab of the dialog, but you will rarely do that. More often, you will want to change the appearance of all variables or a class of variables from the **Settings** menu. From the **Settings** menu, you can change any aspect of the appearance of variables. The **Settings > Variables** menu lets you change the appearance of **All** variables, all **Latent** variables, all **Observed** variables, **Latent Exogenous** variables, **Latent Endogenous** variables, **Observed Exogenous** variables, **Observed Endogenous** variables, **Error** variables, **Latent Error** variables, and **Observed Error** variables.

Draw paths between variables with the tool. Simply click in the source variable's box or oval and drag to the target variable's box or oval. The new path can connect variables either along the line between their center points or at the edge nearest where you click and release. This connection behavior is set in the **Settings > Automation** dialog.

Whenever a path is connected to an exogenous variable, that variable becomes endogenous and an error variable is created. If you do not like the direction of the error variable relative to the newly endogenous variable, use the tool to select the endogenous variable and use the buttons in the Contextual Toolbar to rotate the error variable.

Draw covariances between variables with the tool. If you drag to the right, the curve of the covariance will bend up. If you drag to the left, the curve will bend down. If you drag down, the curve will bend right, and if you drag up, it will bend left. That does not matter greatly; if the curve bends the opposite of what you want, simply click the **Mirror** button in the Contextual Toolbar.

As with variables, you can change more properties of a path or covariance (connection) by double-clicking on or by selecting Properties from the Contextual Toolbar. You can also change the appearance of all connections from the **Settings > Connections** menu. From **Settings > Connections**, you can change the appearance of **All connections**, all **Paths**, all **Covariances**, or **Error Paths**.

Manage variables and connections on the diagram with the select, , tool. As you select objects, their Contextual Toolbars will appear. You can change the name and other properties of selected variables or the properties of selected connections. You can also drag-select or hold the *Shift* key and click to select multiple objects. Changing a property from the Contextual Toolbar or **Property** dialog will affect all the selected objects. You can drag selected variables to other locations on the diagram.

You can place a measurement model (or measurement component of a larger model) on the diagram with the tool. Clicking on the diagram with this tool launches a dialog box where you can name the latent variable, select or type the names of the measurement variables, and specify the direction of the measurements relative to the latent variable. You can even set the spacing between the measurement variables and the distance from the latent variable to the measurements. The tool creates measurement components quickly and with even spacing. Similarly, neatly organized sets of observed variables can be placed with the tool; sets of latent variables, with the tool; and regression components, with the tool.

Place annotations and other text by using the T tool.

We have ignored the locks, , in the Contextual Toolbars. These locks apply constraints to the parameters of the SEM. You can constrain variances, σ^2; means, μ; intercepts, α; and path coefficients and covariances, β. For example, select an exogenous variable or an error variable and type a number in σ^2 to constrain that variance to a fixed value. Or, select three path variables and type a name (a placeholder) in β to constrain all the path coefficients to be equal. You can type numbers, names, or linear expressions in the controls. The linear expressions can involve only numbers and names that are used in other controls.

Do not be afraid to try things. If you do not know what a tool, control, or dialog item does, try it. If you do not like the result, click on the **Undo** button, , in the Standard Toolbar.

Click on in the Standard Toolbar to fit the model. A dialog is launched that allows you to set all the estimation options that are not defined by the path diagram. After estimation, some of the estimation results are displayed on the path diagram. Use the **Results** tab of the **Settings > Variables** and **Settings > Connections** dialogs to change what results are shown and how they appear (font sizes, locations, etc.). Also, as you click on connections or variables, the `Properties Sheet` displays all the estimation results for the selected object.

If you wish to create another model derived from the current model, click in the Standard Toolbar.

Title

lincom — Linear combinations of parameters

Syntax

`sem ..., ...` (fit constrained or unconstrained model)

`lincom` *exp* [, *options*]

Menu

Statistics > Structural equation modeling (SEM) > Testing and CIs > Linear combinations of parameters

Description

`lincom` computes point estimates, standard errors, z statistics, p-values, and confidence intervals for linear combinations of the estimated parameters.

`lincom` is a standard postestimation command and works after `sem` just as it does after any other estimation command except that you must use the `_b[]` coefficient notation; you cannot refer to variables using shortcuts to obtain coefficients on variables.

See [R] **lincom**.

Options

See *Options* in [R] **lincom**.

Remarks

`lincom` works in the metric of SEM, which is to say, path coefficients, variances, and covariances. If you want to frame your linear combinations in terms of standardized coefficients and correlations, prefix `lincom` with `estat stdize:`; see [SEM] **estat stdize**.

Saved results

See *Saved results* in [R] **lincom**.

Also see

[R] **lincom** — Linear combinations of estimators

[SEM] **estat stdize** — Test standardized parameters

[SEM] **nlcom** — Nonlinear combinations of parameters

[SEM] **test** — Wald test of linear hypotheses

Title

lrtest — Likelihood-ratio test of linear hypothesis

Syntax

`sem ..., ...` (fit constrained or unconstrained model)

`estimates store` *modelname1*

`sem ..., ...` (fit unconstrained or constrained model)

`estimates store` *modelname2*

`lrtest` *modelname1 modelname2*

Warning: The two models being compared must include the same observed and latent variables. Place constraints if necessary to achieve your goals. This feature of `lrtest` is unique when `lrtest` is used after `sem`.

Menu

Statistics > Structural equation modeling (SEM) > Testing and CIs > Likelihood-ratio test

Description

`lrtest` performs a likelihood-ratio test comparing two models.

`lrtest` is a standard postestimation command and works after `sem` just as it does after any other estimation command. See [R] **lrtest**.

Remarks

See [SEM] **example 10**.

When using `lrtest` after `sem`, you must be careful that the models being compared have the same observed and latent variables. For instance, the following is allowed:

```
. sem (L1 -> x1 x2 x3) (L1 <- x4 x5) (x1<-x4) (x2<-x5)
. estimates store m1
. sem (L1 -> x1 x2 x3) (L1 <- x4 x5)
. estimates store m2
. lrtest m1 m2
```

The above is allowed because both models have the variables `L1`, `x1`, ..., `x5`.

The following would produce invalid results:

```
. sem (L1 -> x1 x2 x3) (L1 <- x4 x5) (x1<-x4) (x2<-x5)
. estimates store m1
. sem (L1 -> x1 x2 x3) (L1 <- x4)
. estimates store m2
. lrtest m1 m2
```

The second model does not include x5, whereas the first model does. To run this test correctly, you type

```
. sem (L1 -> x1 x2 x3) (L1 <- x4 x5) (x1<-x4) (x2<-x5)
. estimates store m1
. sem (L1 -> x1 x2 x3) (L1 <- x4 x5@0)
. estimates store m2
. lrtest m1 m2
```

Saved results

See *Saved results* in [R] **lrtest**.

Also see

[SEM] **example 10** — MIMIC model

[R] **lrtest** — Likelihood-ratio test after estimation

[SEM] **test** — Wald test of linear hypotheses

[SEM] **estat stdize** — Test standardized parameters

[SEM] **estat eqtest** — Equation-level test that all coefficients are zero

Title

methods and formulas — Methods and formulas

Description

The methods and formulas for the `sem` commands are presented below.

Remarks

Remarks are presented under the following headings:

Variable notation

We will use the following convention to keep track of the five variable types recognized by the `sem` estimation command:

1. Observed endogenous variables are denoted y.
2. Observed exogenous variables are denoted x.
3. Latent endogenous variables are denoted η.
4. Latent exogenous variables are denoted ξ.
5. Error variables are denoted with prefix $e.$ on the associated endogenous variable.
    a. Error variables for observed endogenous are denoted $e.y$.
    b. Error variables for latent endogenous are denoted $e.\eta$.

In any given analysis, there are typically several variables of each type. Vectors of the four main variable types are denoted $\boldsymbol{y}$, $\boldsymbol{x}$, $\boldsymbol{\eta}$, and $\boldsymbol{\xi}$. The vector of all endogenous variables is

$$\boldsymbol{Y} = \begin{pmatrix} \boldsymbol{y} \\ \boldsymbol{\eta} \end{pmatrix}$$

The vector of all exogenous variables is

$$\boldsymbol{X} = \begin{pmatrix} \boldsymbol{x} \\ \boldsymbol{\xi} \end{pmatrix}$$

The vector of all error variables is

$$\boldsymbol{\zeta} = \begin{pmatrix} e.\boldsymbol{y} \\ e.\boldsymbol{\eta} \end{pmatrix}$$

Model and parameterization

`sem` can fit models of the form

$$\boldsymbol{Y} = \mathbf{B}\boldsymbol{Y} + \boldsymbol{\Gamma}\boldsymbol{X} + \boldsymbol{\alpha} + \boldsymbol{\zeta}$$

where $\mathbf{B} = [\beta_{ij}]$ is the matrix of coefficients on endogenous variables predicting other endogenous variables, $\boldsymbol{\Gamma} = [\gamma_{ij}]$ is the matrix of coefficients on exogenous variables, $\boldsymbol{\alpha} = [\alpha_i]$ is the vector of intercepts for the endogenous variables, and $\boldsymbol{\zeta}$ is assumed to have mean 0 and

$$\text{Cov}(\boldsymbol{X}, \boldsymbol{\zeta}) = \mathbf{0}$$

Let

$$\begin{aligned} \boldsymbol{\kappa} &= [\kappa_j] = E(\boldsymbol{X}) \\ \boldsymbol{\Phi} &= [\phi_{ij}] = \text{Var}(\boldsymbol{X}) \\ \boldsymbol{\Psi} &= [\psi_{ij}] = \text{Var}(\boldsymbol{\zeta}) \end{aligned}$$

Then the mean vector of the endogenous variables is

$$\boldsymbol{\mu}_Y = E(\boldsymbol{Y}) = (\boldsymbol{I} - \mathbf{B})^{-1}(\boldsymbol{\Gamma}\boldsymbol{\kappa} + \boldsymbol{\alpha})$$

the variance matrix of the endogenous variables is

$$\boldsymbol{\Sigma}_{YY} = \text{Var}(\boldsymbol{Y}) = (\boldsymbol{I} - \mathbf{B})^{-1}(\boldsymbol{\Gamma}\boldsymbol{\Phi}\boldsymbol{\Gamma}' + \boldsymbol{\Psi}) \left\{(\boldsymbol{I} - \mathbf{B})^{-1}\right\}'$$

and the covariance matrix between the endogenous variables and the exogenous variables is

$$\boldsymbol{\Sigma}_{YX} = \text{Cov}(\boldsymbol{Y}, \boldsymbol{X}) = (\boldsymbol{I} - \mathbf{B})^{-1}\boldsymbol{\Gamma}\boldsymbol{\Phi}$$

Let $\boldsymbol{Z}$ be the vector of all variables:

$$\boldsymbol{Z} = \begin{pmatrix} \boldsymbol{Y} \\ \boldsymbol{X} \end{pmatrix}$$

Then its mean vector is

$$\boldsymbol{\mu} = E(\boldsymbol{Z}) = \begin{pmatrix} \boldsymbol{\mu}_Y \\ \boldsymbol{\kappa} \end{pmatrix}$$

and its variance matrix is

$$\boldsymbol{\Sigma} = \text{Var}(\boldsymbol{Z}) = \begin{pmatrix} \boldsymbol{\Sigma}_{YY} & \boldsymbol{\Sigma}_{YX} \\ \boldsymbol{\Sigma}'_{YX} & \boldsymbol{\Phi} \end{pmatrix}$$

Summary data

Let z_t be the vector of all observed variables for the tth observation,

$$z_t = \begin{pmatrix} y_t \\ x_t \end{pmatrix}$$

and let w_t be the corresponding weight value, where $t = 1, \ldots, N$. If no weights were specified, then $w_t = 1$. Let $w_.$ be the sum of the weights; then the sample mean vector is

$$\overline{z} = \frac{1}{w_.} \sum_{t=1}^{N} w_t z_t$$

and the sample variance matrix is

$$S = \frac{1}{w_. - 1} \sum_{t=1}^{N} w_t (z_t - \overline{z})(z_t - \overline{z})'$$

Maximum likelihood

Let $\boldsymbol{\theta}$ be the vector of unique model parameters, such as

$$\boldsymbol{\theta} = \begin{pmatrix} \text{vec}(\mathbf{B}) \\ \text{vec}(\boldsymbol{\Gamma}) \\ \text{vech}(\boldsymbol{\Psi}) \\ \text{vech}(\boldsymbol{\Phi}) \\ \boldsymbol{\alpha} \\ \boldsymbol{\kappa} \end{pmatrix}$$

Then under the assumption of the multivariate normal distribution, the overall log likelihood for $\boldsymbol{\theta}$ is

$$\log L(\boldsymbol{\theta}) = -\frac{w_.}{2} \left\{ k \log(2\pi) + \log\left\{\det(\boldsymbol{\Sigma}_o)\right\} + \text{tr}\left(\boldsymbol{D}\boldsymbol{\Sigma}_o^{-1}\right) \right\}$$

where k is the number of observed variables, $\boldsymbol{\Sigma}_o$ is the submatrix of $\boldsymbol{\Sigma}$ corresponding to the observed variables, and

$$\boldsymbol{D} = f\boldsymbol{S} + (\overline{z} - \boldsymbol{\mu}_o)(\overline{z} - \boldsymbol{\mu}_o)'$$

where

$$f = \begin{cases} 1, & \text{if } \texttt{nm1} \text{ is specified} \\ \dfrac{w_. - 1}{w_.}, & \text{otherwise} \end{cases}$$

and $\boldsymbol{\mu}_o$ is the subvector of $\boldsymbol{\mu}$ corresponding to the observed variables.

For the BHHH optimization technique and when computing observation-level scores, the log likelihood for $\boldsymbol{\theta}$ is computed as

$$\log L(\boldsymbol{\theta}) = -\sum_{t=1}^{N} \frac{w_t}{2} \left\{ k \log(2\pi) + \log\left\{\det(\boldsymbol{\Sigma}_o)\right\} + (z_t - \boldsymbol{\mu}_o)'\boldsymbol{\Sigma}_o^{-1}(z_t - \boldsymbol{\mu}_o) \right\}$$

and the `nm1` option is ignored.

When `method(mlmv)` is specified, `sem` groups the data according to missing-value patterns. Each missing-value pattern will have its own summary data: k, $\overline{z}$, and $\boldsymbol{S}$. The log likelihood for a missing-value pattern is computed using this summary data and the corresponding elements of $\boldsymbol{\mu}_o$ and $\boldsymbol{\Sigma}_o$. The overall log likelihood is computed by summing the log-likelihood values from each missing-value pattern.

Weighted least squares

Let $\boldsymbol{v}$ be the vector of unique sample moments of the observed variables and $\boldsymbol{\tau}$ be the corresponding vector of population moments. Then

$$\boldsymbol{v} = \begin{bmatrix} \overline{\boldsymbol{z}} \\ \text{vech}(f\boldsymbol{S}) \end{bmatrix}$$

and

$$\boldsymbol{\tau} = \begin{pmatrix} \boldsymbol{\mu}_o \\ \boldsymbol{\Sigma}_o \end{pmatrix}$$

The weighted least squares (WLS) criterion function to minimize is the quadratic form

$$F_{\text{wls}}(\boldsymbol{\theta}) = (\boldsymbol{v} - \boldsymbol{\tau})'\boldsymbol{W}^{-1}(\boldsymbol{v} - \boldsymbol{\tau})$$

where $\boldsymbol{W}$ is the least-squares weight matrix. For unweighted least squares (ULS), the weight matrix is the identity matrix $\boldsymbol{W} = \boldsymbol{I}$. Other weight matrices are mentioned in Bollen (1989).

The weight matrix implemented in `sem` is an estimate of the asymptotic covariance matrix of $\boldsymbol{v}$. This weight matrix is derived without any distributional assumptions and is often referred to as derived from an arbitrary distribution function or is asymptotic distribution free (ADF), thus the option `method(adf)`.

Groups

When the `group()` option is specified, each group has its own summary data and model parameters. The entire collection of model parameters is

$$\boldsymbol{\theta} = \begin{pmatrix} \boldsymbol{\theta}_1 \\ \boldsymbol{\theta}_2 \\ \vdots \\ \boldsymbol{\theta}_G \end{pmatrix}$$

where G is the number of groups. The group-level criterion values are combined to produce an overall criterion value.

For `method(ml)` and `method(mlmv)`, the overall log likelihood is

$$\log L(\boldsymbol{\theta}) = \sum_{g=1}^{G} \log L(\boldsymbol{\theta}_g)$$

For `method(adf)`, the overall criterion function to minimize is

$$F_{\text{wls}}(\boldsymbol{\theta}) = \sum_{g=1}^{G} F_{\text{wls}}(\boldsymbol{\theta}_g)$$

Fitted parameters

sem fits the specified model by maximizing the log likelihood or minimizing the WLS criterion. If $\boldsymbol{\theta}$ is the vector of model parameters, then the fitted parameter vector is denoted by $\widehat{\boldsymbol{\theta}}$, and similarly for $\widehat{\mathbf{B}}$, $\widehat{\boldsymbol{\Gamma}}$, $\widehat{\boldsymbol{\Psi}}$, $\widehat{\boldsymbol{\Phi}}$, $\widehat{\boldsymbol{\alpha}}$, $\widehat{\boldsymbol{\kappa}}$, $\widehat{\boldsymbol{\Sigma}}$, $\widehat{\boldsymbol{\mu}}$, and their individual elements.

Standardized parameters

Let $\widehat{\sigma}_{ii}$ be the ith diagonal element of $\widehat{\boldsymbol{\Sigma}}_{YY}$. Then the standardized parameter estimates are

$$\widetilde{\beta}_{ij} = \widehat{\beta}_{ij}\sqrt{\frac{\widehat{\sigma}_{ii}}{\widehat{\sigma}_{jj}}}$$

$$\widetilde{\gamma}_{ij} = \widehat{\gamma}_{ij}\sqrt{\frac{\widehat{\phi}_{ii}}{\widehat{\sigma}_{jj}}}$$

$$\widetilde{\psi}_{ij} = \begin{cases} \widehat{\psi}_{ij}/\sqrt{\widehat{\psi}_{ii}\widehat{\psi}_{jj}}, & \text{if } i \neq j \\ \widehat{\psi}_{ii}/\sqrt{\widehat{\sigma}_{ii}}, & \text{if } i = j \end{cases}$$

$$\widetilde{\phi}_{ij} = \frac{\widehat{\phi}_{ij}}{\sqrt{\widehat{\phi}_{ii}\widehat{\phi}_{jj}}}$$

$$\widetilde{\alpha}_{i} = \frac{\widehat{\alpha}_{i}}{\sqrt{\widehat{\sigma}_{ii}}}$$

$$\widetilde{\kappa}_{j} = \frac{\widehat{\kappa}_{j}}{\sqrt{\widehat{\phi}_{jj}}}$$

The variance matrix of the standardized parameters is estimated using the delta method.

Reliability

For an observed endogenous variable, y, the reliability may be specified as p or $100 \times p\%$. The variance of e.y is then constrained to $(1 - p)$ times the observed variance of y.

Postestimation

Model framework

estat framework reports the fitted parameters in their individual matrix forms as introduced in *Fitted parameters*.

Goodness of fit

estat gof reports the following goodness-of-fit statistics.

Let the degrees of freedom for the specified model be denoted by df_m. In addition to the specified model, sem fits saturated and baseline models corresponding to the observed variables in the specified model. The saturated model fits a full covariance matrix for the observed variables and has degrees of freedom

$$\mathrm{df}_s = \binom{p+q+1}{2} + p + q$$

where p is the number of observed endogenous variables and q is the number of observed exogenous variables in the model. The baseline model fits a reduced covariance matrix for the observed variables depending on the presence of endogenous variables. If there are no endogenous variables, all variables are uncorrelated in the baseline model; otherwise, only exogenous variables are correlated in the baseline model. The degrees of freedom for the baseline model is

$$\mathrm{df}_b = \begin{cases} 2q, & \text{if } p = 0 \\ 2p + q + \binom{q+1}{2}, & \text{if } p > 0 \end{cases}$$

For method(ml) and method(mlmv), let the saturated log likelihood be denoted by $\log L_s$ and the baseline log likelihood be denoted by $\log L_b$. The likelihood-ratio test of the baseline versus saturated models is computed as

$$\chi^2_{bs} = 2(\log L_s - \log L_b)$$

with degrees of freedom $\mathrm{df}_{bs} = \mathrm{df}_s - \mathrm{df}_b$. The likelihood-ratio test of the specified model versus the saturated model is computed as

$$\chi^2_{ms} = 2\{\log L_s - \log L(\widehat{\boldsymbol\theta})\}$$

with degrees of freedom $\mathrm{df}_{ms} = \mathrm{df}_s - \mathrm{df}_m$.

For method(adf), the saturated criterion value is zero, $F_s = 0$. Let the baseline criterion value be denoted by F_b. The chi-squared test of the baseline versus saturated models is computed as

$$\chi^2_{bs} = NF_b$$

with degrees of freedom $\mathrm{df}_{bs} = \mathrm{df}_s - \mathrm{df}_b$. The chi-squared test of the specified model versus the saturated model is computed as

$$\chi^2_{ms} = NF_{\mathrm{wls}}(\widehat{\boldsymbol\theta})$$

with degrees of freedom $\mathrm{df}_{ms} = \mathrm{df}_s - \mathrm{df}_m$.

The Akaike information criterion (Akaike 1974) is defined as

$$\mathrm{AIC} = -2\log L(\widehat{\boldsymbol\theta}) + 2\mathrm{df}_m$$

The Bayesian information criterion (Schwarz 1978) is defined as

$$\mathrm{BIC} = -2\log L(\widehat{\boldsymbol\theta}) + N\mathrm{df}_m$$

See [R] **BIC note** for additional information on calculating and interpreting BIC.

The overall coefficient of determination is computed as

$$\text{CD} = 1 - \frac{\det(\widehat{\boldsymbol{\Psi}})}{\det(\widehat{\boldsymbol{\Sigma}})}$$

This value is also referred to as the overall R^2 in `estat eqgof` (see [SEM] **estat eqgof**).

The root mean squared error of approximation (Browne and Cudeck 1993) is computed as

$$\text{RMSEA} = \left\{ \frac{(\chi^2_{ms} - \text{df}_{ms})G}{N\text{df}_{ms}} \right\}^{1/2}$$

The 90% confidence interval for RMSEA is

$$\left(\sqrt{\frac{G\lambda_L}{N\text{df}_{ms}}}, \sqrt{\frac{G\lambda_U}{N\text{df}_{ms}}} \right)$$

where λ_L and λ_U are the noncentrality parameters corresponding to a noncentral chi-squared distribution with df_{ms} degrees of freedom in which the noncentral chi-squared random variable has cdf equal to 0.95 and 0.05, respectively.

The Browne and Cudeck (1993) p-value for the test of close fit with null hypothesis,

$$H_0\colon \text{RMSEA} \leq 0.05$$

is computed as

$$p = 1 - \Pr(\chi^2 < \chi^2_{ms} | \lambda, \text{df}_{ms})$$

where χ^2 is distributed noncentral chi-squared with noncentrality parameter $\lambda = (0.05)^2 N\text{df}_{ms}$ and df_{ms} degrees of freedom. This p-value is not computed when the `group()` option is specified.

The comparative fit index (Bentler 1990) is computed as

$$\text{CFI} = 1 - \left[\frac{(\chi^2_{ms} - \text{df}_{ms})}{\max\{(\chi^2_{bs} - \text{df}_{bs}), (\chi^2_{ms} - \text{df}_{ms})\}} \right]$$

The Tucker–Lewis index (Bentler 1990) is computed as

$$\text{TLI} = \frac{(\chi^2_{bs}/\text{df}_{bs}) - (\chi^2_{ms}/\text{df}_{ms})}{(\chi^2_{bs}/\text{df}_{bs}) - 1}$$

Let k be the number of observed variables in the model. If means are not in the fitted model, the standardized root mean squared residual is computed according to Hancock and Mueller (2006)

$$\text{SRMR} = \left\{ \frac{2\sum_{i=1}^{k} \sum_{j \leq i} r^2_{ij}}{k(k+1)G} \right\}^{1/2}$$

where r_{ij} is the standardized covariance residual

$$r_{ij} = \frac{s_{ij}}{\sqrt{s_{ii}s_{jj}}} - \frac{\widehat{\sigma}_{ij}}{\sqrt{\widehat{\sigma}_{ii}\widehat{\sigma}_{jj}}}$$

If means are in the fitted model, SRMR is computed as

$$\text{SRMR} = \left\{ \frac{2\sum_{i=1}^{k}\left(m_i^2 + \sum_{j\le i} r_{ij}^2\right)}{k(k+3)G} \right\}^{1/2}$$

where m_i is the standardized mean residual

$$m_i = \frac{\overline{z}_i}{\sqrt{s_{ii}}} - \frac{\widehat{\mu}_i}{\sqrt{\widehat{\sigma}_{ii}}}$$

These standardized residuals are not the same as those reported by `estat residuals`; see *Residuals* below.

Group goodness of fit

`estat ggof` reports CD, SRMR, and model versus saturated χ^2 values for each group separately. The group-level formulas are the same as those computed for a single group analysis; see *Goodness of fit* above.

Equation-level goodness of fit

`estat eqgof` reports goodness-of-fit statistics for each endogenous variable in the specified model. The coefficient of determination for the ith endogenous variable is computed as

$$R_i^2 = 1 - \frac{\widehat{\psi}_{ii}}{\widehat{\sigma}_{ii}}$$

The Bentler–Raykov (Bentler and Raykov 2000) squared multiple correlation for the ith endogenous variable is computed as

$$\text{mc}_i^2 = \frac{\widehat{\text{Cov}}(y_i, \widehat{y}_i)}{\sqrt{\widehat{\sigma}_{ii}\widehat{\text{Var}}(\widehat{y}_i)}}$$

where $\widehat{\sigma}_{ii}$ is a diagonal element of $\widehat{\boldsymbol{\Sigma}}$, $\widehat{\text{Var}}(\widehat{y}_i)$ is a diagonal element of

$$\widehat{\text{Var}}(\widehat{\boldsymbol{Y}}) = (\boldsymbol{I} - \widehat{\mathbf{B}})^{-1}\widehat{\boldsymbol{\Gamma}}\widehat{\boldsymbol{\Phi}}\widehat{\boldsymbol{\Gamma}}'\left\{(\boldsymbol{I} - \widehat{\mathbf{B}})^{-1}\right\}' + \left\{(\boldsymbol{I} - \widehat{\mathbf{B}})^{-1} - \boldsymbol{I}\right\}\widehat{\boldsymbol{\Psi}}\left\{(\boldsymbol{I} - \widehat{\mathbf{B}})^{-1} - \boldsymbol{I}\right\}'$$

and $\widehat{\text{Cov}}(y_i, \widehat{y}_i)$ is a diagonal element of

$$\widehat{\text{Cov}}(\boldsymbol{Y}, \widehat{\boldsymbol{Y}}) = (\boldsymbol{I} - \widehat{\mathbf{B}})^{-1}\widehat{\boldsymbol{\Gamma}}\widehat{\boldsymbol{\Phi}}\widehat{\boldsymbol{\Gamma}}'\left\{(\boldsymbol{I} - \widehat{\mathbf{B}})^{-1}\right\}' + (\boldsymbol{I} - \widehat{\mathbf{B}})^{-1}\widehat{\boldsymbol{\Psi}}\left\{(\boldsymbol{I} - \widehat{\mathbf{B}})^{-1} - \boldsymbol{I}\right\}'$$

Wald tests

`estat eqtest` performs Wald tests on the coefficients for each endogenous equation in the model. `estat ginvariant` computes a Wald test of group invariance for each model parameter that is free to vary across all groups. See [R] **test**.

Score tests

`estat mindices` computes modification indices for each constrained parameter in the model, including paths and covariances that were not even part of the model specification. Modification indices are score tests, which are also known as Lagrange multiplier tests. `estat scoretests` performs a score test for each user-specified linear constraint. `estat ginvariant` performs a score test of group invariance for each model parameter that is constrained to be equal across all groups.

A score test compares a constrained model fit to the same model without one or more constraints. The score test is computed as

$$\chi^2 = \boldsymbol{g}(\widehat{\boldsymbol{\theta}})'\boldsymbol{V}(\widehat{\boldsymbol{\theta}})\boldsymbol{g}(\widehat{\boldsymbol{\theta}})$$

where $\widehat{\boldsymbol{\theta}}$ is the fitted parameter vector from the constrained model, $\boldsymbol{g}(\boldsymbol{\theta})$ is the gradient vector function for the unconstrained model, and $\boldsymbol{V}(\boldsymbol{\theta})$ is the variance matrix function computed from the expected information matrix function for the unconstrained model. For `method(ml)` and `method(mlmv)`,

$$\begin{aligned} \boldsymbol{g}(\boldsymbol{\theta}) &= \frac{\partial \log L^*(\boldsymbol{\theta})}{\partial \boldsymbol{\theta}} \\ \boldsymbol{V}(\boldsymbol{\theta}) &= \left[E\left\{ -\frac{\partial^2 \log L^*(\boldsymbol{\theta})}{\partial \boldsymbol{\theta} \partial \boldsymbol{\theta}'} \right\} \right]^{-1} \end{aligned}$$

where $\log L^*(\boldsymbol{\theta})$ is the log-likelihood function for the unconstrained model. For `method(adf)`,

$$\begin{aligned} \boldsymbol{g}(\boldsymbol{\theta}) &= -\frac{\partial F^*_{\text{wls}}(\boldsymbol{\theta})}{\partial \boldsymbol{\theta}} \\ \boldsymbol{V}(\boldsymbol{\theta}) &= \left[E\left\{ \frac{\partial^2 F^*_{\text{wls}}(\boldsymbol{\theta})}{\partial \boldsymbol{\theta} \partial \boldsymbol{\theta}'} \right\} \right]^{-1} \end{aligned}$$

where $F^*(\boldsymbol{\theta})$ is the WLS criterion function for the unconstrained model.

The score test is computed as described in Wooldridge (2010) when `vce(robust)` or `vce(cluster` *clustvar*`)` is specified.

Residuals

`estat residuals` reports raw, normalized, and standardized residuals for means and covariances of the observed variables.

The raw residual for the mean of the ith observed variable is

$$\overline{z}_i - \widehat{\mu}_i$$

The raw residual for the covariance between the ith and jth observed variables is

$$S_{ij} - \widehat{\sigma}_{ij}$$

The normalized residual for the mean of the ith observed variable is

$$\frac{\overline{z}_i - \widehat{\mu}_i}{\sqrt{\widehat{\text{Var}}(\overline{z}_i)}}$$

where

$$\widehat{\text{Var}}(\overline{z}_i) = \begin{cases} \dfrac{S_{ii}}{N}, & \text{if the sample option is specified} \\ \dfrac{\widehat{\sigma}_{ii}}{N}, & \text{otherwise} \end{cases}$$

The normalized residual for the covariance between the ith and jth observed variables is

$$\frac{S_{ij} - \widehat{\sigma}_{ij}}{\sqrt{\widehat{\text{Var}}(S_{ij})}}$$

where

$$\widehat{\text{Var}}(S_{ij}) = \begin{cases} \dfrac{S_{ii}S_{jj} + S_{ij}}{N}, & \text{if the sample option is specified} \\ \dfrac{\widehat{\sigma}_{ii}\widehat{\sigma}_{jj} + \widehat{\sigma}_{ij}}{N}, & \text{otherwise} \end{cases}$$

If the `nm1` option is specified, the denominator in the variance estimates is $N - 1$ instead of N.

The standardized residual for the mean of the ith observed variable is

$$\frac{\overline{z}_i - \widehat{\mu}_i}{\sqrt{\widehat{\text{Var}}(\overline{z}_i - \widehat{\mu}_i)}}$$

where

$$\widehat{\text{Var}}(\overline{z}_i - \widehat{\mu}_i) = \widehat{\text{Var}}(\overline{z}_i) - \widehat{\text{Var}}(\widehat{\mu}_i)$$

and $\widehat{\text{Var}}(\widehat{\mu}_i)$ is computed using the delta method. Missing values are reported when the computed value of $\widehat{\text{Var}}(\overline{z}_i)$ is less than $\widehat{\text{Var}}(\widehat{\mu}_i)$. The standardized residual for the covariance between the ith and jth observed variables is

$$\frac{S_{ij} - \widehat{\sigma}_{ij}}{\sqrt{\widehat{\text{Var}}(S_{ij} - \widehat{\sigma}_{ij})}}$$

where

$$\widehat{\text{Var}}(S_{ij} - \widehat{\sigma}_{ij}) = \widehat{\text{Var}}(S_{ij}) - \widehat{\text{Var}}(\widehat{\sigma}_{ij})$$

and $\widehat{\text{Var}}(\widehat{\sigma}_{ij})$ is computed using the delta method. Missing values are reported when the computed value of $\widehat{\text{Var}}(S_{ij})$ is less than $\widehat{\text{Var}}(\widehat{\sigma}_{ij})$. The variances of the raw residuals used in the standardized residual calculations are derived in Hausman (1978).

Testing standardized parameters

`estat stdize` provides access to tests on the standardized parameter estimates. `estat stdize` can be used as a prefix to `lincom` (see [R] **lincom**), `nlcom` (see [R] **nlcom**), `test` (see [R] **test**), and `testnl` (see [R] **testnl**).

Stability of nonrecursive systems

`estat stable` reports a stability index for nonrecursive systems. The stability index is calculated as the maximum of the modulus of the eigenvalues of $\mathbf{B}$. The nonrecursive system is considered stable if the stability index is less than one.

Direct, indirect, and total effects

`estat teffects` reports direct, indirect, and total effects for the fitted model. The direct effects are

$$\boldsymbol{E}_d = \begin{bmatrix} \widehat{\mathbf{B}} & \widehat{\boldsymbol{\Gamma}} \end{bmatrix}$$

the total effects are

$$\boldsymbol{E}_t = \begin{bmatrix} (\boldsymbol{I} - \widehat{\mathbf{B}})^{-1} - \boldsymbol{I} & , & \widehat{\boldsymbol{\Gamma}}(\boldsymbol{I} - \widehat{\mathbf{B}})^{-1}\boldsymbol{\Gamma} \end{bmatrix}$$

and the indirect effects are $\boldsymbol{E}_i = \boldsymbol{E}_t - \boldsymbol{E}_d$. The standard errors of the effects are computed using the delta method.

Let $\boldsymbol{D}$ be the diagonal matrix whose elements are the square roots of the diagonal elements of $\widehat{\boldsymbol{\Sigma}}$, and let $\boldsymbol{D}_Y$ be the submatrix of $\boldsymbol{D}$ associated with the endogenous variables. Then the standardized effects are

$$\widetilde{\boldsymbol{E}}_d = \boldsymbol{D}_Y^{-1}\boldsymbol{E}_d\boldsymbol{D}$$
$$\widetilde{\boldsymbol{E}}_i = \boldsymbol{D}_Y^{-1}\boldsymbol{E}_i\boldsymbol{D}$$
$$\widetilde{\boldsymbol{E}}_t = \boldsymbol{D}_Y^{-1}\boldsymbol{E}_t\boldsymbol{D}$$

Predictions

`predict` computes factor scores and linear predictions.

Factor scores are computed with a linear regression using the mean vector and variance matrix from the fitted model. For notational convenience, let

$$\boldsymbol{Z} = \begin{pmatrix} \boldsymbol{z} \\ \boldsymbol{l} \end{pmatrix}$$

where

$$\boldsymbol{z} = \begin{pmatrix} \boldsymbol{y} \\ \boldsymbol{x} \end{pmatrix}$$

and

$$\boldsymbol{l} = \begin{pmatrix} \boldsymbol{\eta} \\ \boldsymbol{\xi} \end{pmatrix}$$

The fitted mean of $\boldsymbol{Z}$ is

$$\widehat{\boldsymbol{\mu}}_Z = \begin{pmatrix} \widehat{\boldsymbol{\mu}}_z \\ \widehat{\boldsymbol{\mu}}_l \end{pmatrix}$$

and fitted variance of $\boldsymbol{Z}$ is

$$\widehat{\boldsymbol{\Sigma}}_Z = \begin{pmatrix} \widehat{\boldsymbol{\Sigma}}_{zz} & \widehat{\boldsymbol{\Sigma}}_{zl} \\ \widehat{\boldsymbol{\Sigma}}_{zl}' & \widehat{\boldsymbol{\Sigma}}_{ll} \end{pmatrix}$$

The factor scores are computed as

$$\widetilde{\boldsymbol{l}} = \begin{pmatrix} \widetilde{\boldsymbol{\eta}} \\ \widetilde{\boldsymbol{\xi}} \end{pmatrix} = \widehat{\boldsymbol{\Sigma}}_{zl}'\widehat{\boldsymbol{\Sigma}}_{zz}\widehat{\boldsymbol{\mu}}_z + \widehat{\boldsymbol{\mu}}_l$$

The linear prediction for the endogenous variables in the tth observation is computed as

$$\widehat{\boldsymbol{Y}}_t = \widehat{\mathbf{B}}\widetilde{\boldsymbol{Y}}_t + \widehat{\boldsymbol{\Gamma}}\widetilde{\boldsymbol{X}}_t + \widehat{\boldsymbol{\alpha}}$$

where

$$\widetilde{\boldsymbol{Y}}_t = \begin{pmatrix} \boldsymbol{y}_t \\ \widetilde{\boldsymbol{\eta}} \end{pmatrix}$$

and

$$\widetilde{\boldsymbol{X}}_t = \begin{pmatrix} \boldsymbol{x}_t \\ \widetilde{\boldsymbol{\xi}} \end{pmatrix}$$

Also see

[SEM] **sem** — Structural equation model estimation command

Title

nlcom — Nonlinear combinations of parameters

Syntax

`sem ..., ...` (fit constrained or unconstrained model)

`nlcom` *exp* [, *options*]

Menu

Statistics > Structural equation modeling (SEM) > Testing and CIs > Nonlinear combinations of parameters

Description

`nlcom` computes point estimates, standard errors, z statistics, p-values, and confidence intervals for (possibly) nonlinear combinations of the estimated parameters.

`nlcom` is a standard postestimation command and works after `sem` just as it does after any other estimation command.

See [R] **nlcom**.

Options

See *Options* in [R] **nlcom**.

Remarks

`nlcom` works in the metric of SEM, which is to say, path coefficients, variances, and covariances. If you want to frame your nonlinear combinations in terms of standardized coefficients and correlations, prefix `nlcom` with `estat stdize:`; see [SEM] **estat stdize**.

❑ Technical note

`estat stdize:` is, strictly speaking, unnecessary because everywhere you wanted a standardized coefficient or correlation, you could just type the formula. If you did that, you would get the same results but for numerical precision. The answer produced with the `estat stdize:` prefix will be a little more accurate because `estat stdize:` is able to substitute an analytic derivative in one part of the calculation where `nlcom`, doing the whole thing itself, would be forced to use a numeric derivative.

❑

Saved results

See *Saved results* in [R] **nlcom**.

Also see

[R] **nlcom** — Nonlinear combinations of estimators

[SEM] **estat stdize** — Test standardized parameters

[SEM] **lincom** — Linear combinations of parameters

[SEM] **test** — Wald test of linear hypotheses

Title

predict — Factor scores, linear predictions, etc.

Syntax

```
sem ..., ...                          (fit constrained or unconstrained model)

predict [type] {stub*|newvarlist} [if] [in] [, options]
```

options	Description
xb	linear prediction for all OEn variables; the default
xb(*varlist*)	linear prediction for specified OEn variables
xblatent	linear prediction for all LEn variables
xblatent(*varlist*)	linear prediction for specified LEn variables
latent	factor scores for all latent variables
latent(*varlist*)	factor scores for specified latent variables
scores	calculate first derivative of the log likelihood

Key: OEn = observed endogenous; LEn = latent endogenous

Menu

Statistics > Structural equation modeling (SEM) > Predictions

Description

predict creates new variables containing observation-by-observation values of estimated factor scores (meaning predicted values of latent variables) and predicted values for latent and observed endogenous variables. Out-of-sample prediction is allowed.

When predict is used on a model fit by sem with the group() option, results are produced using the appropriate group-specific estimates. Out-of-sample prediction is allowed; missing values are filled in for groups not included at the time the model was fit.

predict allows two syntaxes. You can type

```
. predict stub*, ...
```

to create variables named *stub*1, *stub*2, ..., or you can type

```
. predict var1 var2 ..., ...
```

to create variables named *var1*, *var2*, ...

predict may not be used with summary statistics data.

Options

`xb` calculates the linear prediction for all observed endogenous variables in the model. `xb` is the default if no option is specified.

`xb(`*varlist*`)` calculates the linear prediction for the variables specified, all of which must be observed endogenous variables.

`xblatent` and `xblatent(`*varlist*`)` calculate the linear prediction for all or the specified latent endogenous variables, respectively.

`latent` and `latent(`*varlist*`)` calculate the factor scores for all or the specified latent variables, respectively. The calculation method is an analogue of regression scoring; namely, it produces the means of the latent variables conditional on the observed variables used in the model. If missing values are found among the observed variables, conditioning is on the variables with observed values only.

`scores` is for use by programmers. It provides the first derivative of the observation-level log likelihood with respect to the parameters.

Programmers: In single-group `sem`, each parameter that is not constrained to be 0 has an associated equation. As a consequence, the number of equations, and hence the number of score variables created by `predict`, may be large.

Remarks

See [SEM] **example 14**.

Factor scoring for latent variables can be interpreted as a form of missing-value imputation—think of each latent variable as an observed variable that has only missing values.

When latent variables are present in the model, linear predictions from `predict, xb` are computed by substituting the factor scores in place of each latent variable before computing the linear combination of coefficients. This method will lead to inconsistent coefficient estimates when the factor score contains measurement error; see Bollen (1989, 305–306).

Also see

[SEM] **example 14** — Predicted values

[SEM] **methods and formulas** — Methods and formulas

[SEM] **sem postestimation** — Postestimation tools for sem

Title

sem — Structural equation model estimation command

Syntax

`sem` *paths* [*if*] [*in*] [*weight*] [, *options*]

where *paths* are the paths of the model in command-language path notation; see [SEM] **sem path notation**.

options	Description
model_description_options	fully define, along with *paths*, the model to be fit
group_options	fit model for different groups
ssd_options	for use with summary statistics data
estimation_options	method used to obtain estimation results
reporting_options	reporting of estimation results
syntax_options	controlling interpretation of syntax

`bootstrap`, `by`, `jackknife`, `permute`, `statsby`, and `svy` are allowed; see [U] **11.1.10 Prefix commands**.
`fweight`s, `iweight`s, and `pweight`s are allowed; see [U] **11.1.6 weight**.
Also see [SEM] **sem postestimation** for features used after model estimation.

Menu

Statistics > Structural equation modeling (SEM) > Model building and estimation

Description

`sem` fits structural equation models. Even when you use the GUI, you are using the `sem` command.

Options

model_description_options describe the model to be fit. The model to be fit is fully specified by *paths*—which appear immediately after `sem`—and the options `covariance()`, `variance()`, and `means()`. See [SEM] **sem model description options** and [SEM] **sem path notation**.

group_options allow the specified model to be fit for different subgroups of the data, with some parameters free to vary across groups and other parameters constrained to be equal across groups. See [SEM] **sem group options**.

ssd_options allow models to be fit using summary statistics data (SSD), meaning data on means, variances (standard deviations), and covariances (correlations). See [SEM] **sem ssd options**.

estimation_options control how the estimation results are obtained. These options control how the standard errors (VCE) are obtained and control technical issues such as choice of estimation method. See [SEM] **sem estimation options**.

reporting_options control how the results of estimation are displayed. See [SEM] **sem reporting options**.

syntax_options control how the syntax that you type is interpreted. See [SEM] **sem syntax options**.

Remarks

For a readable explanation of what `sem` can do and how to use it, see any of the intro sections. You might start with [SEM] **intro 1**.

For examples of `sem` in action, see any of the example sections. You might start with [SEM] **example 1**.

For detailed syntax and descriptions, see the references below.

Remarks on three advanced topics are presented under the following headings:

Default normalization constraints
Default covariance assumptions
How to solve convergence problems

Default normalization constraints

`sem` applies the following rules as necessary to identify the model:

1. `means(1: LatentExogenous@0)`
   `sem` constrains all latent exogenous variables to have mean 0. When the `group()` option is specified, this rule is applied to the first group only.
2. `(LatentEndogenous <- _cons@0)`
   `sem` sets all latent endogenous variables to have intercept 0.
3. `(first <- Latent@1)`
   If latent variable `Latent` is measured by observed endogenous variables, then `sem` sets the path coefficient of `(first<-Latent)` to be 1; `first` is the first observed endogenous variable.
4. `(First<-Latent@1)`
   If (3) does not apply—if latent variable `Latent` is measured by other latent endogenous variables only—`sem` sets the path coefficient of `First<-Latent` to be 1; `First` is the first latent variable.

The above constraints are applied as needed. Here are the available overrides:

1. To override the normalization constraints, specify your own constraints. Most normalization constraints are added by `sem` as needed. See *How sem solves the problem for you* under *Identification 2: Normalization constraints (anchoring)* in [SEM] **intro 3**.
2. To override `means()` constraints, you must use the `means()` option to free the parameter. To override that the mean of latent exogenous variable `MyLatent` has mean 0, specify the `means(MyLatent)` option. See [SEM] **sem path notation**.
3. To override constrained path coefficients from `_cons`, such as `(LatentEndogenous <- _cons@0)`, you must explicitly specify the path without a constraint `(LatentEndogenous <- _cons)`. See [SEM] **sem path notation**.

Default covariance assumptions

`sem` assumes the following covariance structure:

1. `covstructure(_Ex, unstructured)`
   All exogenous variables (observed and latent) are assumed to be correlated with each other.

2. `covstructure(e._En, diagonal)`
   The error variables associated with all endogenous variables are assumed to be uncorrelated with each other.

You can override these assumptions by

1. Constraining or specifying the relevant covariance to allow `e.y` and `e.Ly` to be correlated (specify the `covariance(e.y*e.Ly)` option); see [SEM] **sem model description options**.

2. Using the `covstructure()` option; see [SEM] **sem option covstructure( )**.

How to solve convergence problems

Structural equation models often have difficulty converging. We more than touched on this in [SEM] **intro 3**. If you experience convergence difficulties, we offer the following advice.

1. If you are specifying `sem`'s `reliability()` option, remove it and try fitting the model again. If the model converges, then your estimate of the reliability is too low; see *What can go wrong* in [SEM] **sem option reliability( )**.

2. Be sure to let `sem` provide its default normalization constraints. By default, `sem` (1) constrains all latent exogenous variables to have mean 0; (2) constrains all latent endogenous variables to have intercept 0; and (3) constrains the paths from latent variables (endogenous or exogenous) to the first observed endogenous variable to have coefficient 1.

   Replacing any of the above defaults can cause problems, but problems are most likely to arise if you replace default (3). Do not constrain path coefficients merely to obtain model identification. Let `sem` choose those constraints.

   Attempt to fit your model again. If it converges, you have solved your problem if you are willing to accept results with the default identifying constraints. Those results, we emphasize, are mathematically equivalent to results with any other set of identifying constraints. If you must have results with your identifying constraints, you will need to reimpose your constraints one at a time and find good starting values for the other parameters that lead to convergence.

3. Check whether your model has any feedback loops, such as

   ```
   . sem ... (y1<-y2 x2) (y2<-y1 x3) ...
   ```

   In this example, variable `y1` affects `y2` affects `y1`. Models with such feedback loops are said to be nonrecursive. Assume you had a solution to the above model. The results might be unstable in a substantive sense; see *nonrecursive (structural) model (system)* in [SEM] **Glossary**. The problem is that finding such truly unstable solutions is often difficult and the stability problem manifests itself as a convergence problem.

   Understand, if you have convergence problems and you have feedback loops, that is not proof that the underlying values are unstable.

   To check for this problem, temporarily remove the feedback loop,

   ```
   . sem ... (y1<-y2 x2) (y2<-   x3) ...
   ```

and see whether the model converges. If it does, then the true coefficients are probably unstable. You need to put back the feedback loop and provide reasonable starting values. Based on the model that did converge, you now have good starting values for the path `y1<-y2` and for `y2<-x3`. Say those fitted values were 1.52 and 2.73. Now fit the model:

```
. sem ... (y1<-(y2, init(1.52)) x2) (y2<-y1 (x3, init(2.73))) ...
```

Although we show specifying the starting values using the standard path notation, using the parameters from the simplified model as starting values is even easier; see [SEM] **sem option from()**.

That model may very well converge whether the solution is stable or not. If it converges, check for stability using `estat stable`. If it does not converge, try initializing the `y1<-y2` coefficient to 0. After that, you simply have to try different values. You know the model without the feedback loop can be fit and that solution can be used to provide good starting values. You now must find which variables need to have starting values specified. Best is to use `sem` option `from()`.

4. At this point, you have dismissed problems (1), (2), and (3) as the culprits. You now need to follow a strategy of temporarily simplifying your model and fitting the simplified model. Once you find a simplified model that does converge, you can use the fitted values as starting values as you reintroduce the complication you removed; see [SEM] **sem option from()**. Perhaps you reintroduce the complication all at once, or perhaps reintroduce it piece by piece, getting better and better starting values along the way.

   If your model has a measurement component, we recommend focusing your attention on that part first. Simplify your model by temporarily deleting the rest of it.

   Additional guidance can be found in *Starting values* in [SEM] **intro 3** and in [SEM] **sem option from()**.

Saved results

`sem` saves the following in `e()`:

Scalars

`e(N)`	number of observations
`e(N_clust)`	number of clusters
`e(N_groups)`	number of groups
`e(N_missing)`	number of missing values in the sample for `method(mlmv)`
`e(ll)`	log likelihood of model
`e(df_m)`	model degrees of freedom
`e(df_b)`	baseline model degrees of freedom
`e(df_s)`	saturated model degrees of freedom
`e(chi2_ms)`	test of target model against saturated model
`e(df_ms)`	degrees of freedom for `e(chi2_ms)`
`e(p_ms)`	p-value for `e(chi2_ms)`
`e(chi2_bs)`	test of baseline model against saturated model
`e(df_bs)`	degrees of freedom for `e(chi2_bs)`
`e(p_bs)`	p-value for `e(chi2_bs)`
`e(rank)`	rank of `e(V)`
`e(ic)`	number of iterations
`e(rc)`	return code
`e(converged)`	1 if target model converged, 0 otherwise
`e(critvalue)`	log likelihood or discrepancy of fitted model
`e(critvalue_b)`	log likelihood or discrepancy of baseline model
`e(critvalue_s)`	log likelihood or discrepancy of saturated model
`e(modelmeans)`	1 if fitting means and intercepts, 0 otherwise

Macros

e(cmd)	sem
e(cmdline)	command as typed
e(data)	raw or ssd if SSD data was used
e(wtype)	weight type
e(wexp)	weight expression
e(title)	title in estimation output
e(clustvar)	name of cluster variable
e(vce)	vcetype specified in vce()
e(vcetype)	title used to label Std. Err.
e(method)	estimation method: ml, mlmv, or adf
e(technique)	maximization technique
e(properties)	b V
e(estat_cmd)	program used to implement estat
e(predict)	program used to implement predict
e(lyvars)	names of latent y variables
e(oyvars)	names of observed y variables
e(lxvars)	names of latent x variables
e(oxvars)	names of observed x variables
e(groupvar)	name of group variable
e(xconditional)	empty if noxconditional specified, xconditional otherwise

Matrices

e(b)	parameter vector
e(b_std)	standardized parameter vector
e(b_pclass)	parameter class
e(V)	covariance matrix of the estimators
e(V_std)	standardized covariance matrix of the estimators
e(V_modelbased)	model-based variance
e(admissible)	admissibility of Σ, Ψ, Φ
e(ilog)	iteration log (up to 20 iterations)
e(gradient)	gradient vector
e(nobs)	vector with number of observations per group
e(groupvalue)	vector of group values of e(groupvar)
e(S[_#])	sample covariance matrix of observed variables (for group #)
e(means[_#])	sample means of observed variables (for group #)
e(W)	weight matrix for method(adf)

Functions

e(sample)	marks estimation sample (not with summary statistics data)

Also see

[SEM] **intro 1** — Introduction

[SEM] **sem path notation** — Command syntax for path diagrams

[SEM] **sem model description options** — Model description options

[SEM] **sem group options** — Fitting models on different groups

[SEM] **sem ssd options** — Options for use with summary statistics data

[SEM] **sem estimation options** — Options affecting estimation

[SEM] **sem reporting options** — Options affecting reporting of results

[SEM] **sem syntax options** — Options affecting interpretation of syntax

[SEM] **sem postestimation** — Postestimation tools for sem

[SEM] **methods and formulas** — Methods and formulas

Title

sem estimation options — Options affecting estimation

Syntax

`sem` *paths* ..., ... *estimation_options*

estimation_options	Description
`method()`	estimation method; see [SEM] **sem option method()**
`vce()`	VCE type; see [SEM] **sem option method()**
`nm1`	compute sample variance rather than ML variance
`noxconditional`	compute covariances, etc., of observed exogenous variables
`allmissing`	for use with `method(mlmv)`
`noivstart`	skip calculation of starting values
maximize_options	control maximization process for specified model; seldom used
`satopts(`*maximize_options*`)`	control maximization process for saturated model; seldom used
`baseopts(`*maximize_options*`)`	control maximization process for baseline model; seldom used

Description

These options control how results are obtained.

Options

`method()` and `vce()` specify the method used to obtain parameter estimates and the technique used to obtain the variance–covariance matrix of the estimates. See [SEM] **sem option method()**.

`nm1` specifies that the variances and covariances used in the SEM equations be the sample variances (divided by $N - 1$) and not the asymptotic variances (divided by N). This is a minor technical issue of little importance unless you are trying to match results from other software that assumes sample variances. `sem` assumes asymptotic variances.

`noxconditional` states that you wish to include the means, variances, and covariances of the observed exogenous variables among the parameters to be estimated by `sem`. See [SEM] **sem option noxconditional**.

`allmissing` specifies how missing values should be treated when `method(mlmv)` is also specified.

Usually, `sem` omits from the estimation sample observations that contain missing values of any of the observed variables used in the model. `method(mlmv)`, however, can deal with these missing values, and in that case, observations containing missing are not omitted.

Even so, `sem, method(mlmv)` does omit observations containing `.a`, `.b`, ..., `.z` from the estimation sample. `sem` assumes you do not want these observations used, because the missing value is not missing at random. If you wish `sem` to include these observations in the estimation sample, specify the `allmissing` option.

`noivstart` is an arcane option that is of most use to programmers. It specifies that `sem` is to skip efforts to produce good starting values using instrumental-variable techniques, techniques that require computer time. If you specify this option, you should specify all the starting values. Any starting values not specified will be assumed to be 0 (means, path coefficients, and covariances) or some simple function of the data (variances).

maximize_options specify the standard and rarely specified options for controlling the maximization process; see [R] **maximize**. The relevant options for `sem` are `difficult`, `technique(`*algorithm_spec*`)`, `iterate(`*#*`)`, `[no]log`, `trace`, `gradient`, `showstep`, `hessian`, `tolerance(`*#*`)`, `ltolerance(`*#*`)`, and `nrtolerance(`*#*`)`.

`satopts(`*maximize_options*`)` is a rarely specified option and is only relevant if you specify the `method(mlmv)` option. `sem` reports a test for model versus saturated at the bottom of the output. Thus `sem` needs to obtain the saturated fit. In the case of `method(ml)` or `method(adf)`, `sem` can make a direct calculation. In the other case of `method(mlmv)`, `sem` must actually fit the saturated model. The maximization options specified inside `satopts()` control that maximization process. It is rare that you need to specify the `satopts()` option even if you find it necessary to specify the overall *maximize_options*.

`baseopts(`*maximize_options*`)` is a rarely specified option and an irrelevant one unless you also specify `method(mlmv)` or `method(adf)`. When fitting the model, `sem` records information about the baseline model for later use by `estat gof`, should you use that command. Thus `sem` needs to obtain the baseline fit. In the case of `method(ml)`, `sem` can make a direct calculation. In the cases of `method(mlmv)` and `method(adf)`, `sem` must actually fit the baseline model. The maximization options specified inside `baseopts()` control that maximization process. It is rare that you need to specify the `baseopts()` option even if you find it necessary to specify the overall *maximize_options*.

Remarks

The most commonly specified option among this group is `vce()`. See [SEM] **intro 7**.

Also see

[SEM] **sem** — Structural equation model estimation command

[SEM] **sem option method()** — Specifying method and calculation of VCE

[SEM] **sem option noxconditional** — Computing means, etc. of observed exogenous variables

[SEM] **intro 7** — Robust and clustered standard errors

Title

sem group options — Fitting models on different groups

Syntax

`sem` *paths* ..., ... *group_options*

group_options	Description
`group(`*varname*`)`	fit model for different groups
`ginvariant(`*classname*`)`	specify parameters that are equal across groups

classname	Description
`scoef`	structural coefficients
`scons`	structural intercepts
`mcoef`	measurement coefficients
`mcons`	measurement intercepts
`serrvar`	covariances of structural errors
`merrvar`	covariances of measurement errors
`smerrcov`	covariances between structural and measurement errors
`meanex`	means of exogenous variables
`covex`	covariances of exogenous variables
`all`	all of the above
`none`	none of the above

`ginvariant(mcoef mcons)` is the default if `ginvariant()` is not specified.

`meanex`, `covex`, and `all` exclude the observed exogenous variables (that is, they include only the latent exogenous variables) unless you specify the `noxconditional` option or the `noxconditional` option is otherwise implied; see [SEM] **sem option noxconditional**. This is what you would desire in most cases.

Description

`sem` can fit combined models across subgroups of the data and allow some parameters to vary and constrain others to be equal across subgroups. These subgroups could be males and females, age category, and the like.

`sem` performs such estimation when the `group(`*varname*`)` option is specified. The `ginvariant(`*classname*`)` option specifies which parameters are to be constrained to be equal across the groups.

Options

`group(`*varname*`)` specifies that the model be fit as described above. *varname* specifies the name of a numeric variable that records the group to which the observation belongs.

If you are using summary statistics data in place of raw data, *varname* is the name of the group variable as reported by `ssd describe`; see [SEM] **ssd**.

`ginvariant(`*classname*`)` specifies which classes of parameters of the model are to be constrained to be equal across groups. The classes are defined above. The default is `ginvariant(mcoef mcons)` if the option is not specified.

Remarks

See [SEM] **intro 5**, and see [SEM] **example 20** and [SEM] **example 23**.

Also see

[SEM] **sem** — Structural equation model estimation command

[SEM] **intro 5** — Comparing groups

[SEM] **example 20** — Two-factor measurement model by group

[SEM] **example 23** — Specifying parameter constraints across groups

Title

sem model description options — Model description options

Syntax

`sem` *paths* ..., ... *model_description_options*

model_description_options	Description
* covariance()	path notation for treatment of covariances; see [SEM] **sem path notation**
* variance()	path notation for treatment of variances; see [SEM] **sem path notation**
* means()	path notation for treatment of means; see [SEM] **sem path notation**
* covstructure()	alternative method to place restrictions on covariances; see [SEM] **sem option covstructure()**
noconstant	do not fit intercepts
nomeans	do not fit means or intercepts
noanchor	do not apply default anchoring
forcenoanchor	programmer's option
* reliability()	reliability of measurement variables; see [SEM] **sem option reliability()**
constraints()	specify constraints; see [SEM] **sem option constraints()**
from()	specify starting values; see [SEM] **sem option from()**

* Option may be specified more than once.

Description

paths and the options above describe the model to be fit.

Options

`covariance()`, `variance()`, and `means()` fully describe the model to be fit. See [SEM] **sem path notation**.

`covstructure()` provides a convenient way to constrain covariances in your model. Alternatively or in combination, you can place constraints using the standard path notation. See [SEM] **sem option covstructure()**.

`noconstant` specifies that all intercepts be constrained to zero. See [SEM] **sem path notation**.

`nomeans` specifies that means and intercepts not be fit. The means and intercepts are concentrated out of the function being optimized, which function is typically the likelihood function. Results for all other parameters are the same whether or not this option is specified.

This option is seldom specified. `sem` issues this option to itself when you use summary statistics data that do not include summary statistics for the means.

`noanchor` specifies that `sem` is not to check for lack of identification and fill in anchors where needed. `sem` is instead to issue an error message if anchors would be needed. You specify this option when you believe you have specified the necessary normalization constraints and if you are wrong, want to hear about it. See *Identification 2: Normalization constraints (anchoring)* in [SEM] **intro 3**.

`forcenoanchor` is similar to `noanchor` except that rather than issue an error message, `sem` proceeds to estimation. There is no reason you should specify this option. `forcenoanchor` is used in testing of `sem` at StataCorp.

`reliability()` specifies the fraction of variance not due to measurement error for a variable. See [SEM] **sem option reliability( )**.

`constraints()` specifies parameter constraints you wish to impose on your model; see [SEM] **sem option constraints( )**. Constraints can also be specified as described in [SEM] **sem path notation**, and they are usually more conveniently specified using the path notation.

`from()` specifies the starting values to be used in the optimization process; see [SEM] **sem option from( )**. Starting values can also be specified by using the `init()` suboption as described in [SEM] **sem path notation**.

Remarks

To use `sem` successfully, you need to understand *paths*, `covariance()`, `variance()`, and `means()`; see *Using path diagrams to specify the model* in [SEM] **intro 2** and [SEM] **sem path notation**.

`covstructure()` is often convenient; see [SEM] **sem option covstructure( )**.

Also see

[SEM] **sem** — Structural equation model estimation command

[SEM] **intro 2** — Learning the language: Path diagrams and command language

[SEM] **sem path notation** — Command syntax for path diagrams

[SEM] **sem option covstructure( )** — Specifying covariance restrictions

[SEM] **sem option reliability( )** — Fraction of variance not due to measurement error

[SEM] **sem option constraints( )** — Specifying constraints

[SEM] **sem option from( )** — Specifying starting values

Title

sem option constraints() — Specifying constraints

Syntax

```
sem ... [, ... constraints(# [# ...]) ...]
```

where # are constraint numbers. Constraints are defined by the `constraint` command; see [R] **constraint**.

Description

Constraints refer to constraints to be imposed on the estimated parameters of a model. These constraints usually come in one of three forms:

1. Constraints that a parameter such as a path coefficient or variance is equal to a fixed value such as 1.
2. Constraints that two or more parameters are equal.
3. Constraints that two or more parameters are related by a linear equation.

It is usually easier to specify constraints using `sem`'s path notation; see [SEM] **sem path notation**.

`sem`'s `constraints()` option provides an alternative way of specifying constraints.

Remarks

There is only one case where `constraints()` might be easier to use than specifying constraints in the path notation. You wish to specify that two or more parameters are related, and then decide you would like to fix the value at which they are related.

For example, if you wanted to specify that parameters are equal, you could type

```
. sem ... (y1<-x@c1) (y2<-x@c1)      (y3<-x@c1)        ...
```

Using the path notation, you can specify more general relationships, too, such as

```
. sem ... (y1<-x@c1) (y2<-x@(2*c1) (y3<-x@(3*c1+1)) ...
```

Say you now decide you want to fix *c1* at value 1. Using the path notation, you modify what you previously typed:

```
. sem ... (y1<-x@1) (y2<-x@2)      (y3<-x@4)        ...
```

Alternatively, you could do the following:

```
. constraint 1 _b[y2:x] = 2*_b[y1:x]
. constraint 2 _b[y3:x] = 3*_b[y1:x] + 1
. sem ..., ... constraints(1 2)
. constraint 3 _b[y1:x] = 1
. sem .., ... constraints(1 2 3)
```

See [R] **constraint**.

Also see

[SEM] **sem** — Structural equation model estimation command

[SEM] **sem path notation** — Command syntax for path diagrams

[SEM] **sem model description options** — Model description options

[R] **constraint** — Define and list constraints

Title

sem option covstructure( ) — Specifying covariance restrictions

Syntax

sem ... [, ... covstructure(*variables*, *structure*) ...]

sem ... [, ... covstructure(*groupid*: *variables*, *structure*) ...]

where *variables* is one of

1. a list of (a subset of the) exogenous variables in your model, for example,

   ```
   . sem ..., ... covstruct(x1 x2, structure)
   ```

2. _OEx, meaning all observed exogenous variables in your model
3. _LEx, meaning all latent exogenous variables in your model
4. _Ex, meaning all exogenous variables in your model

or where *variables* is one of

1. a list of (a subset of the) error variables in your model, for example,

   ```
   . sem ..., ... covstruct(e.y1 e.y2 e.Aspect, structure)
   ```

2. e._OEn, meaning all error variables associated with observed endogenous variables in your model
3. e._LEn, meaning all error variables associated with latent endogenous variables in your model
4. e._En, meaning all error variables in your model

and where *structure* is

structure	Description	Notes
`diagonal`	all variances unrestricted all covariances fixed at 0	
`unstructured`	all variances unrestricted all covariances unrestricted	
`identity`	all variances equal all covariances fixed at 0	
`exchangeable`	all variances equal all covariances equal	
`zero`	all variances fixed at 0 all covariances fixed at 0	
`pattern(`*matname*`)`	covariances (variances) unrestricted if *matname*$[i,j] \geq .$ covariances (variances) equal if *matname*$[i,j] =$ *matname*$[k,l]$	(1)
`fixed(`*matname*`)`	covariances (variances) unrestricted if *matname*$[i,j] \geq .$ covariances (variances) fixed at *matname*$[i,j]$ otherwise	(2)

Notes:

(1) Only elements in the lower triangle of *matname* are used. All values in *matname* are interpreted as the `floor()` of the value if noninteger values appear. Row and column stripes of *matname* are ignored.

(2) Only elements on the lower triangle of *matname* are used. Row and column stripes of *matname* are ignored.

groupid may be specified only when the `group()` option is also specified, and even then it is optional; see [SEM] **sem group options**.

Description

`sem` option `covstructure()` provides a convenient way to constrain the covariances of your model.

Alternatively or in combination, you can place constraints on the covariances using the standard path notation, such as

```
. sem ..., ... cov(name1*name2@c1 name3*name4@c1) ...
```

See [SEM] **sem path notation**.

Option

`covstruct(`[*groupid*`:`] *variables*`,` *structure*`)` is used either (1) to modify the covariance structure among the exogenous variables of your model or (2) to modify the covariance structure among the error variables of your model.

You may specify the `covstruct()` option multiple times.

The default covariance structure for the exogenous variables is `covstruct(_Ex, unstructured)`.

The default covariance structure for the error variables is `covstruct(e._En, diagonal)`.

Remarks

See [SEM] **example 17**.

SEM allows covariances among exogenous variables, both latent and observed, and allows covariances among the error variables. Covariances between exogenous variables and error variables are disallowed (assumed to be 0).

Some authors refer to the covariances among the exogenous variables as matrix $\mathbf{\Phi}$ and to the covariances among the error variables as matrix $\mathbf{\Psi}$.

Also see

[SEM] **sem** — Structural equation model estimation command

[SEM] **sem path notation** — Command syntax for path diagrams

[SEM] **example 17** — Correlated uniqueness model

Title

sem option from() — Specifying starting values

Syntax

`sem ...` [`, ... from(`*matname*[`, skip`]`) ...`]

`sem ...` [`, ... from(`*svlist*`) ...`]

where *matname* is the name of a Stata matrix and

where *svlist* is a space-separated list of the form

eqname`:`*name* = #

Description

See [SEM] **intro 3** for a description of starting values.

Starting values are usually not specified. When there are convergence problems, it is often necessary to specify starting values. You can specify starting values

1. using suboption `init()` as described in [SEM] **sem path notation**, or
2. using `sem` option `from()` described here.

`sem` option `from()` is especially useful when you use the solution of one model as starting values for another.

Option

`skip` is an option of `from(`*matname*`)`. It specifies that parameters in *matname* that do not appear in the model being fit be skipped. If this option is not specified, the existence of such parameters causes `sem` to issue an error message. This option is rarely specified. Usually, *matname* contains a subset, not a superset, of the values being estimated.

Remarks

Remarks are presented under the following headings:

Syntax 1, especially useful when dealing with convergence problems

Say you are attempting to fit

`. sem` *your_full_model*`, ...`

and are having difficulty with convergence. Following the advice in *How to solve convergence problems* in [SEM] **sem**, you have simplified your model,

```
. sem your_simple_model, ...
```

and that does converge. In the advice section, you are told to now use those starting values in your full model. Let's imagine that there are 47 estimated parameters in *your_simple_model*.

Using the standard `init()` method for specifying starting values, you now have a real job in front of you. You have to type your full model, find all the places where you now have starting values, and add an `init()` suboption. Just a piece of your full model might read

```
... (y<-L1 L2) (L1->x1 x2) (L2->x3 L4) ...
```

and you need to modify that to read

```
... (y<-(L1, init(14.283984)) L2)                        ///
    (L1->(x1, init(2.666532)) (x2, init(-6.39499)))      ///
    (L2->x3 L4) ...
```

That change handles just 3 of the 47 parameters you need to specify.

There is an easier way. Type

```
. sem your_simple_model, ...
. matrix b = e(b)
. sem your_full_model, ... from(b)
```

Here is how this works:

1. You fit the simple model. `sem` stores the resulting parameters in `e(b)`.
2. You store the fitted parameters in Stata matrix `b`.
3. You now fit your full model, typing the model just as you would usually, and you add option `from(b)`. That option tells `sem` to get any starting values it can from Stata matrix `b`. `sem` gets all the starting values it can from `b` and then follows its usual logic for producing starting values for the remaining parameters.

Just because you use the `from(b)` option does not mean you cannot specify starting values the usual way for other parameters. You can even specify starting values for some of the same parameters. Starting values specified by suboption `init()` take precedence over those obtained from `from()`.

Syntax 2, seldom used

In syntax 2, you specify

```
. sem ..., ... from(eqname:name=# eqname:name=# ...)
```

For instance, you could type

```
. sem ..., ... from(var(X):_cons=10)
```

or you could type

```
. sem ..., ... var(X, init(10))
```

It is usually easier to type the second. See [SEM] **sem path notation**.

You may combine the two notations. If starting values are specified for a parameter both ways, those specified by `init()` take precedence.

Also see

[SEM] **sem** — Structural equation model estimation command

[SEM] **sem path notation** — Command syntax for path diagrams

[SEM] **sem model description options** — Model description options

[R] **maximize** — Details of iterative maximization

Title

sem option method() — Specifying method and calculation of VCE

Syntax

sem ... [, ... method(*method*) vce(*vcetype*) ...]

method	Description
ml	maximum likelihood; the default
mlmv	ml with missing values
adf	asymptotic distribution free

vcetype	Description
oim	observed information matrix; the default
eim	expected information matrix
opg	outer product of gradients
robust	Huber/White/sandwich estimator
cluster *clustvar*	generalized Huber/White/sandwich estimator
bootstrap [, *bootstrap_options*]	bootstrap estimation
jackknife [, *jackknife_options*]	jackknife estimation

The following combinations of method() and vce() are allowed:

	oim	eim	opg	robust	cluster	bootstrap	jackknife
ml	x	x	x	x	x	x	x
mlmv	x	x	x	x	x	x	x
adf	x	x				x	x

Description

sem option method() specifies the method used to obtain the estimated parameters.

sem option vce() specifies the technique used to obtain the variance–covariance matrix of the estimates (VCE), which includes the reported standard errors.

Options

method(*method*) specifies the method used to obtain parameter estimates. method(ml) is the default.

vce(*vcetype*) specifies the technique used to obtain the VCE. vce(oim) is the default.

Remarks

See

1. *Assumptions and choice of estimation method* in [SEM] **intro 3**
2. [SEM] **intro 7**
3. [SEM] **intro 8**

Also see

[SEM] **sem** — Structural equation model estimation command

[SEM] **intro 3** — Substantive concepts

[SEM] **intro 7** — Robust and clustered standard errors

[SEM] **intro 8** — Standard errors, the full story

[SEM] **example 26** — Fitting a model using data missing at random

Title

sem option noxconditional — Computing means, etc. of observed exogenous variables

Syntax

```
sem ... [, ... noxconditional ...]
```

Description

`sem` has a `noxconditional` option that you may rarely wish to specify. The option is described below.

Option

`noxconditional` states that you wish to include the means, variances, and covariances of the observed exogenous variables among the parameters to be estimated by `sem`.

Remarks

Remarks are presented under the following headings:

What is x conditional?

In many cases, `sem` does not include the means, variances, and covariances of observed exogenous variables among the parameters to be estimated. When `sem` omits them, the estimator of the model is said to be x conditional. Rather than estimating the values of the means, variances, and covariances, `sem` uses the separately calculated observed values of those statistics. `sem` does this to save time and memory.

`sem` does not use the x-conditional calculation when it would be inappropriate.

The `noxconditional` option prevents `sem` from using the x-conditional calculation. You specify `noxconditional` on the `sem` command:

```
. sem ..., ... noxconditional
```

Do not confuse the x-conditional calculation with the assumption of conditional normality discussed in *Conditional normality might be sufficient* in [SEM] **intro 3**. The x-conditional calculation is appropriate even when the assumption of conditional normality is inappropriate.

When to specify noxconditional

It is never inappropriate to specify the noxconditional option. Be aware, however,

1. If you are using the default method(ml), estimated point estimates and standard errors will be the same.
2. If you are using method(adf), estimated point estimates and standard errors will be slightly different, asymptotically equivalent, and there is no reason to prefer one set of estimates over the other.
3. If you are using method(mlmv), the situation is the same as in (1).
4. Regardless of the estimation method used, calculation of results will require more computer time and memory. The memory requirements increase quadratically with the total number of estimated parameters in your model. If you have k_1 observed exogenous variables and k_2 latent exogenous variables, the number of added parameters from noxconditional is $k_1 + k_1(k_1 + 1)/2 + k_1 k_2$. The resulting total memory requirements can be so great as to require more memory than your computer can provide.

To make statements (1) to (4) true, there are two cases when sem specifies noxconditional for you:

1. sem defaults to noxconditional whenever you constrain a mean, variance, or covariance of an observed exogenous variable. For example,

```
. sem ..., ... means(x1@m x2@m)
. sem ..., ... var(x1@v x2@v)
. sem ..., ... cov(x1*x2@c x1*x3@c)
. sem ..., ... covstruct(_OEx, diagonal)
```

   See [SEM] **sem path notation** and [SEM] **sem option covstructure( )**.

2. sem defaults to noxconditional whenever you use method(mlmv) and there are missing values among the observed exogenous variables.

There are only two reasons for you to specify the noxconditional option:

1. Specify noxconditional if you subsequently wish to test means, variances, or covariances of observed exogenous variables with postestimation commands. For example,

```
. sem ..., ... noxconditional
. sem, coeflegend
. test _b[means(x1):_cons] == _b[means(x2)_cons]
```

2. Specify noxconditional if you are fitting a model using the group() option.
3. You also specify the ginvariant() option, and you want the ginvariant() classes meanex, covex, or all to include the observed exogenous variables. For example,

```
. sem ..., ... by(agegrp) ginvariant(all) noxconditional
```

You may also wish to specify noxconditional when comparing results with those from other packages. Many packages use the noxconditional approach when using an estimation method other than maximum likelihood (ML). Correspondingly, most packages use the x-conditional calculation when using ML.

Option forcexconditional (a technical note)

In addition to `noxconditional`, `sem` has a `forcexconditional` option:

`sem ...` [, ... `forcexconditional` ...]

This option turns off `sem`'s switching away from the x-conditional calculation when that is required. Do not specify this option unless you are exploring the behavior of x-conditional calculation in cases where it is theoretically inappropriate.

Also see

[SEM] **sem** — Structural equation model estimation command

Title

sem option reliability( ) — Fraction of variance not due to measurement error

Syntax

```
sem ... [, ... reliability(varname # [varname # [...]])]
```

where *varname* is the name of an observed endogenous variable and # is the fraction or percentage of variance not due to measurement error:

```
. sem ..., ... reliability(x1 .8  x2 .9)
. sem ..., ... reliability(x1 80%  x2 90%)
```

Description

`sem` option `reliability()` allows you to specify the fraction of variance not due to measurement error for measurement variables.

Option

`reliability(`*varname* # [...]`)` specifies the reliability for variable *varname*. Reliability is bounded by 0 and 1 and is equal to

$$\frac{1 - \text{noise variance}}{\text{total variance}}$$

The reliability is assumed to be 1 when not specified.

Remarks

See [SEM] **example 24**.

Remarks are presented under the following headings:

Background
Dealing with measurement error of exogenous variables
Dealing with measurement error of endogenous variables
What can go wrong

Background

Variables measured with error have attenuated path coefficients. If we had the model

```
. sem (y<-x)
```

and `x` were measured with error, then the estimated path coefficient would be biased toward zero. The usual solution to such measurement problems is to find multiple measurements and develop a latent variable from them:

```
. sem (x1 x2 x3<-X) (y<-X)
```

Another solution is available if we know the reliability of x. In that case, we can fit the model

```
. sem (x<-X) (y<-X), reliability(x .9)
```

The two solutions can even be combined:

```
. sem (x1 x2 x3<-X) (y<-X), reliability(x1 .9  x2 .8  x3 .9)
```

Even if you do not know the reliability, you can experiment using different but reasonable values for the reliability and thus determine the sensitivity of your estimation results to the measurement problem.

Dealing with measurement error of exogenous variables

Measurement error is most important when it occurs in exogenous variables, yet the `reliability()` option deals with measurement error of endogenous variables only. By creation of a latent variable, `reliability()` can deal with the measurement error of exogenous variables.

To fit the model `(y<-x)` where `x` is measured with error, you must introduce a latent variable corresponding to `x` measured without error. That is, the model `(y<-x)` can be converted into the model `(x<-X)` and `(y<-X)`:

$$x = \alpha_0 + \beta_0 X + e.x$$

$$y = \alpha_1 + \beta_1 X + e.y$$

To fit this model, you type

```
. sem (x<-X) (y<-X), reliability(x .9)
```

`sem` will introduce a normalization constraint, namely, that the path coefficient β_0 for `x<-X` is 1, but that is of no importance. What is important is that the estimate of that path coefficient β_1 of `y<-X` is the coefficient that would be obtained from `y<-x` were `x` measured without error.

In the above, we specified the measurement part of the model first. Be sure to do that. You might think you could equally well reverse the two terms so that, rather than writing

`(x<-X) (y<-X)`   *(correct)*

you could write

`(y<-X) (x<-X)`   *(incorrect)*

But you cannot do that unless you write

`(y<-X) (x<-X@1)`   *(correct)*

because otherwise results are as if you typed

`(y<-X@1) (x<-X)`   *(incorrect)*

All of that is because `sem` places its normalization constraint from the latent variable to the first observed endogenous variable. There is no real error if the terms are interchanged except that you will be surprised by the coefficient of 1 for `y<-X` and (the reciprocal of) the coefficient of interest will be on `x<-X`.

See *How sem solves the problem for you* in [SEM] **intro 3** and see *Default normalization constraints* in [SEM] **sem**.

Dealing with measurement error of endogenous variables

When a variable would already be endogenous before you add the `reliability()` option, it really makes little difference whether you add the `reliability()` option. That is because endogenous variables are assumed to contain error, and if some of that error is measurement error, it is still just an error. Coefficients will be unchanged by the inclusion of the `reliability()` option.

Some variances and covariances will change, but the changes are offsetting in the calculation of path coefficients.

What will change are the standardized coefficients should you ask to see them. That is because the variances are changed.

What can go wrong

Consider a model of `y` on `x`. Say we fit the model using linear regression. If the R^2 of the fit is 0.6, then we know the reliability must be greater than 0.6. R^2 measures the fraction of variance of `y` that is explained by `x`, and the reliability of `x` measures the fraction of the variance of `x` that is not due to measurement error. Measurement error is assumed to be pure noise. It is just not possible that we could explain 0.6 of the variance of `y` using a variable with reliability of, say, 0.5.

Well, in fact, it is because there is always a chance that the pure noise will correlate with `y`, too. Asymptotically, that probability vanishes, but in finite—especially small—samples, it could happen. Even so, the calculation of the corrected SEM estimates blows up.

If you have convergence problems, you need to check for this. Specify a larger value for the reliability. The problem is, you cannot specify a value of 1, and large values such as 0.99999 can lead to a lack of identification. In most cases, you will be able to find a value in between, but the only way to be sure is to remove the `reliability()` option and, if necessary, simplify your model by removing any intermediary latent variables you had to add because of reliability.

If your model converges without reliability, then your measure of reliability is too low. At this point, we have little useful advice for you. Check whether you have the right value, of course. If you do, then there are two possibilities: either the experts who provided that estimate are wrong or you got unlucky in that the measurement error did just happen to correlate with the rest of your data. You will need to evaluate the chances of that for yourself. In any case, you can experiment with higher values of the reliability and at least provide an idea of the sensitivity of your estimates to differing assumptions.

Also see

[SEM] **sem** — Structural equation model estimation command

[SEM] **sem model description options** — Model description options

[SEM] **example 24** — Reliability

Title

sem option select( ) — Using sem with summary statistics data

Syntax

```
sem ... [, ... select(# [# ...]) ... ]
```

Description

sem may be used with summary statistics data (SSD), data containing only summary statistics such as the means, standard deviations or variances, and correlations and covariances of the underlying, raw data.

You enter SSD using the `ssd` command; see [SEM] **ssd**.

To fit a model with `sem`, there is nothing special you have to do except specify the `select()` option where you would usually specify `if` *exp*.

Option

`select(#` [*#* ...]`)` is allowed only when you have SSD in memory. It specifies which groups should be used.

Remarks

See [SEM] **intro 10**, entitled *Fitting models using summary statistics data.*

`sem` option `select()` is the SSD alternative for `if` *exp* if only you had the underlying, raw data in memory. With the underlying raw data, where you would usually type

```
. sem ... if agegrp==1 | agegrp==3, ...
```

with SSD in memory, you type

```
. sem ..., ... select(1 3)
```

You may select only groups for which you have separate summary statistics recorded in your summary statistics dataset; the `ssd describe` command will list the group variable, if any. See [SEM] **ssd**.

By the way, `select()` may be combined with `sem` option `group()`. Where you might usually type

```
. sem ... if agegrp==1 | agegrp==3, ... group(agegrp)
```

you type

```
. sem ..., ... select(1 3) group(agegrp)
```

The above restricts `sem` to age groups 1 and 3, so the result will be an estimation of a combined model of age groups 1 and 3 with some coefficients allowed to vary between the groups and other coefficients constrained to be equal across the groups. See [SEM] **sem group options**.

Also see

[SEM] **sem** — Structural equation model estimation command

[SEM] **intro 10** — Fitting models using summary statistics data

Title

sem path notation — Command syntax for path diagrams

Syntax

sem *paths* ... [, covariance() variance() means() [group()]]

paths specifies the direct paths between the variables of your model.

The model to be fit is fully described by *paths*, covariance(), variance(), and means().

The syntax of these elements is modified (generalized) when the group() option is specified.

Description

The command syntax for describing your structural equation models is fully specified by *paths*, covariance(), variance(), and means(). How this works is described below.

Options

covariance() is used to

1. specify that a particular covariance path of your model that usually is assumed to be 0 be estimated,
2. specify that a particular covariance path that usually is assumed to be nonzero is not to be estimated (to be constrained to be 0),
3. constrain a covariance path to a fixed value, such as 0, 0.5, 1, etc., and
4. constrain two or more covariance paths to be equal.

variance() does the same as covariance() except it does it with variances.

means() does the same as covariance() except it does it with means.

group() is mentioned here only because the syntax of *paths* and the arguments of covariance(), variance(), and means() gains an extra syntactical piece when group() is specified.

Remarks

Remarks are presented under the following headings:

Model notation when option group() is not specified

Path notation is used by the sem command to specify the model to be fit, for example,

```
. sem (x1 x2 x3 x4 <- X)
. sem (L1 -> x1 x2 x3 x4 x5) (L2 -> x6 x7 x8 x9 x10)
```

In the path notation,

1. Latent variables are indicated by a *name* in which at least the first letter is capitalized.
2. Observed variables are indicated by a *name* in which at least the first letter is lowercased. Observed variables correspond to variable names in the dataset.
3. Error variables, while mathematically a special case of latent variables, are considered in a class by themselves. Every endogenous variable (whether observed or latent) automatically has an error variable associated with it. The error variable associated with endogenous variable *name* is `e.`*name*.
4. Paths between variables are written as

    ```
    (name1 <- name2)
    ```

    or

    ```
    (name2 -> name1)
    ```

    There is no significance to which coding is used.
5. Paths between the same variables can be combined: The paths

    ```
    (name1 <- name2) (name1 <- name3)
    ```

    can be combined as

    ```
    (name1 <- name2 name3)
    ```

    or as

    ```
    (name2 name3 -> name1)
    ```

    The paths

    ```
    (name1 <- name3) (name2 <- name3)
    ```

    can be combined as

    ```
    (name1 name2 <- name3)
    ```

    or as

    ```
    (name3 -> name1 name2)
    ```

    The paths

    ```
    (name1 <- name2 name3)
    (name4 <- name2 name3)
    ```

    may be written as

    ```
    (name1 name4 <- name2 name3)
    ```

    or as

    ```
    (name2 name3 -> name1 name4)
    ```

6. Variances and covariances (curved paths) between variables are indicated by options. Variances are indicated by

    ```
    ..., ... var(name1)
    ```

    Covariances are indicated by

    ```
    ..., ... cov(name1*name2)
    ..., ... cov(name2*name1)
    ```

There is no significance to the order of the names.

The actual names of the options are `variance()` and `covariance()`, but they are invariably abbreviated as `var()` and `cov()`, respectively.

The `var()` and `cov()` options are the same option, so a variance can be typed as

`..., ... cov(`*name1*`)`

and a covariance can be typed as

`..., ... var(`*name1*`*`*name2*`)`

7. Variances may be combined, covariances may be combined, and variances and covariances may be combined.

   If you have

   `..., ... var(`*name1*`) var(`*name2*`)`

   you may code this as

   `..., ... var(`*name1 name2*`)`

   If you have

   `..., ... cov(`*name1*`*`*name2*`) cov(`*name2*`*`*name3*`)`

   you may code this as

   `..., ... cov(`*name1*`*`*name2 name2*`*`*name3*`)`

   All the above combined can be coded as

   `..., ... var(`*name1 name2 name1*`*`*name2 name2*`*`*name3*`)`

   or as

   `..., ... cov(`*name1 name2 name1*`*`*name2 name2*`*`*name3*`)`

8. All variables except endogenous variables are assumed to have a variance; it is only necessary to code the `var()` option if you wish to place a constraint on the variance or specify an initial value. See items 11, 12, 13, and 16 below.

   Endogenous variables have a variance, of course, but that is the variance implied by the model. If *name* is an endogenous variable, then `var(`*name*`)` is invalid. The error variance of the endogenous variable is `var(e.`*name*`)`.

9. Variables mostly default to being correlated:

   a. All exogenous variables are assumed to be correlated with each other, whether observed or latent.

   b. Endogenous variables are never directly correlated, although their associated error variables can be.

   c. All error variables are assumed to be uncorrelated with each other.

   You can override these defaults on a variable-by-variable basis using the `cov()` option.

   To assert that two variables are uncorrelated that otherwise would be assumed to be correlated, constrain the covariance to be 0:

   `..., ... cov(`*name1*`*`*name2*`@0)`

   To allow two variables to be correlated that otherwise would be assumed to be uncorrelated, simply specify the existence of the covariance:

   `..., ... cov(`*name1*`*`*name2*`)`

This latter is especially commonly done with errors:

```
..., .. cov(e.name1*e.name2)
```

10. Means of variables are indicated by option:

```
..., ... means(name)
```

Variables mostly default to having nonzero means:

a. All observed exogenous variables are assumed to have nonzero means. The means can be constrained using the `means()` option, but only if you are performing `noxconditional` estimation; see [SEM] **sem option noxconditional**.

b. Latent exogenous variables are assumed to have mean 0. Means of latent variables are not estimated by default. If you specify enough normalization constraints to identify the mean of a latent exogenous variable, you can specify `means(`*Name*`)` to indicate that the mean should be estimated.

c. Endogenous variables have no separate mean. Their means are those implied by the model. The `means()` option may not be used with endogenous variables.

d. Error variables have mean 0 and this cannot be modified. The `means()` option may not be used with error variables.

To constrain the mean to a fixed value, such as 57, code

```
..., ... means(name@57)
```

Separate `means()` options may be combined:

```
..., ... means(name1@57 name2@100)
```

11. Fixed-value constraints may be specified for a path, variance, covariance, or mean by using `@` (the at sign). For example,

```
(name1 <- name2@1)

(name1 <- name2@1 name3@1)

..., ... var(name@100)

..., ... cov(name1*name2@223)

..., ... cov(name1@1 name2@1 name1*name2@.8)

..., ... means(name@57)
```

12. Symbolic constraints may be specified for a path, variance, covariance, or mean by using `@` (the at sign). For example,

```
(name1 <- name2@c1) (name3 <- name4@c1)

..., ... var(name1@c1 name2@c1)

..., ... cov(name1@1 name2@1 name3@1 name1*name2@c1 name1*name3@c1)

..., ... means(name1@c1 name2@c1)

(name1 <- name2@c1) ..., var(name3@c1) means(name4@c1)
```

Symbolic names are just names from 1 to 32 characters in length. Symbolic constraints constrain equality. For simplicity, all constraints below will have names `c1`, `c2`, ...

13. Linear combinations of symbolic constraints may be specified for a path, variance, covariance, or mean by using `@` (the at sign). For example,

    (*name1* <- *name2*@c1) (*name3* <- *name4*@(2*c1))

    ..., ... var(*name1*@c1 *name2*@(c1/2))

    ..., ... cov(*name1*@1 *name2*@1 *name3*@1 *name1***name2*@c1 *name1***name2*@(c1/2))

    ..., ... means(*name1*@c1 *name2*@(3*c1+10))

    (*name1* <- *name2*@(c1/2)) ..., var(*name3*@c1) means(*name4*@(2*c1))

14. All equations in the model are assumed to have an intercept (to include observed exogenous variable `_cons`) unless the `noconstant` option (abbreviation `nocon`) is specified, and then all equations are assumed not to have an intercept (not to include `_cons`).

    Regardless of whether `noconstant` is specified, you may explicitly refer to observed exogenous variable `_cons`.

    The following path specifications are ways of writing the same model:

    (*name1* <- *name2*) (*name1* <- *name3*)

    (*name1* <- *name2*) (*name1* <- *name3*) (*name1* <- _cons)

    (*name1* <- *name2 name3*)

    (*name1* <- *name2 name3* _cons)

    There is no reason to explicitly specify `_cons` unless (1) you have also specified the `noconstant` option and want to include `_cons` in some equations but not others or (2) regardless of whether you specified the `noconstant` option, you wish to place a constraint on its path coefficient. For example,

    (*name1* <- *name2 name3* _cons@c1) (*name4* <- *name5* _cons@c1)

15. The `noconstant` option may be specified globally or within a path specification. That is,

    (*name1* <- *name2 name3*) (*name4* <- *name5*), nocon

    suppresses the intercepts in both equations. Alternatively,

    (*name1* <- *name2 name3*, nocon) (*name4* <- *name5*)

    suppresses the intercept in the first equation but not the second, whereas

    (*name1* <- *name2 name3*) (*name4* <- *name5*, nocon)

    suppresses the intercept in the second equation but not the first.

    In addition, consider the equation

    (*name1* <- *name2 name3*, nocon)

    This can be written equivalently as

    (*name1* <- *name2*, nocon) (*name1* <- *name3*, nocon)

16. Initial values (starting values) may be specified for a path, variance, covariance, or mean by using the `init(`*#*`)` suboption:

    (*name1* <- (*name2*, init(0)))

    (*name1* <- (*name2*, init(0)) *name3*)

    (*name1* <- (*name2*, init(0)) (*name3*, init(5)))

    ..., ... var((*name3*, init(1)))

```
..., ... cov((name4*name5, init(.5)))
..., ... means((name5, init(0)))
```

The initial values may be combined with symbolic constraints:

```
(name1 <- (name2@c1, init(0)))
(name1 <- (name2@c1, init(0)) name3)
(name1 <- (name2@c1, init(0)) (name3@c2, init(5)))
..., ... var((name3@c1, init(1)))
..., ... cov((name4*name5@c1, init(.5)))
..., ... means((name5@c1, init(0)))
```

The above fully describes *paths* and the arguments of options `means()`, `variance()`, and `covariance()` in the case when the `group()` option is not specified.

Added syntax when option group() is specified

The model you wish to fit is fully described by the *paths*, `covariance()`, `variance()`, and `means()` that you type.

The `group(`*varname*`)` option,

```
. sem ..., ... group(varname)
```

specifies that the model be fit separately for the different values of *varname*. *varname* might be `sex` and then the model would be fit separately for males and females, or *varname* might be something else and perhaps take on more than two values.

Whatever *varname* is, `group(`*varname*`)` defaults to letting some of the path coefficients, covariances, variances, and means of your model vary across the groups and constrains others to be equal. Which parameters vary and which are constrained is described in [SEM] **sem group options**, but that is a minor detail right now.

In what follows, we will assume that *varname* is `mygrp` and takes on three values. Those values are 1, 2, and 3, but they could just as well be 2, 9, and 12.

Consider typing

```
. sem ..., ...
```

and typing

```
. sem ..., ... group(mygrp)
```

Whatever the *paths*, `covariance()`, `variance()`, and `means()` are that describe the model, there are now three times as many parameters because each group has its own unique set. In fact, when you give the second command, you are not merely asking for three times the parameters, you are specifying three models, one for each group! In this case, you specified the same model three times without knowing it.

You can vary the model specified across groups.

1. Let's write the model you wish to fit as

```
. sem (a) (b) (c), cov(d) cov(e) var(f)
```

where *a*, *b*, ..., *f* stand for what you type. In this generic example, we have two `cov()` options just because multiple `cov()` options often occur in real models. When you type

```
. sem (a) (b) (c), cov(d) cov(e) var(f) group(mygrp)
```

results are as if you typed

```
. sem (1: a) (2: a) (3: a)                                  ///
      (1: b) (2: b) (3: b)                                  ///
      (1: c) (2: c) (3: c),                                 ///
              cov(1: d) cov(2: d) cov(3: d)                 ///
              cov(1: e) cov(2: e) cov(3: e)                 ///
              var(1: f) cov(2: f) cov(3: f)   group(mygrp)
```

The `1:`, `2:`, and `3:` identify the groups for which paths, covariances, or variances are being added, modified, or constrained.

If `mygrp` contained the unique values 5, 8, and 10 instead of 1, 2, and 3, then `5:` would appear in place of `1:`; `8:` would appear in place of `2:`; and `10:` would appear in place of `3:`.

2. Consider the model

```
. sem (y <- x) (b) (c), cov(d) cov(e) var(f) group(mygrp)
```

If you wanted to constrain the path coefficient `(y <- x)` to be the same across all three groups, you could type

```
. sem (y <- x@c1) (b) (c), cov(d) cov(e) var(f) group(mygrp)
```

See item 12 above for more examples of specifying constraints. This works because the expansion of `(y <- x@c1)` is

```
(1: y <- x@c1) (2: y <- x@c1) (3: y <- x@c1)
```

3. Consider the model

```
. sem (y <- x) (b) (c), cov(d) cov(e) var(f) group(mygrp)
```

If you wanted to constrain the path coefficient `(y <- x)` to be the same in groups 2 and 3, you could type

```
. sem (1: y <- x) (2: y <- x@c1) (3: y <- x@c1) (b) (c),        ///
                                  cov(d) cov(e) var(f) group(mygrp)
```

4. Instead of following item 3, you could type

```
. sem (y <- x) (2: y <- x@c1) (3: y <- x@c1) (b) (c),          ///
                                  cov(d) cov(e) var(f) group(mygrp)
```

The part `(y <- x) (2: y <- x@c1) (3: y <- x@c1)` expands to

```
(1: y <- x) (2: y <- x) (3: y <- x) (2: y <- x@c1) (3: y <- x@c1)
```

and thus the path is defined twice for group 2 and twice for group 3. When a path is defined more than once, the definitions are combined. In this case, the second definition adds more information, so the result is as if you typed

```
(1: y <- x) (2: y <- x@c1) (3: y <- x@c1)
```

5. Instead of following item 3 or item 4, you could type

```
. sem (y <- x@c1) (1: y <- x@c2) (b) (c),        ///
                              cov(d) cov(e) var(f) group(mygrp)
```

The part (y <- x@c1) (1: y <- x@c2) expands to

```
(1: y <- x@c1)  (2: y <- x@c1)  (3: y <- x@c1)  (1: y <- x@c2)
```

When results are combined from repeated definitions, definitions that appear later take precedence. In this case, results are as if the expansion read

```
(1: y <- x@c2) (2: y <- x@c1)  (3: y <- x@c1)
```

Thus coefficients for groups 2 and 3 are constrained. The group-1 coefficient is constrained to *c2*. If *c2* appears nowhere else in the model specification, then results are as if the path for group 1 were unconstrained.

6. Instead of following item 3, item 4, or item 5, you could not type

```
. sem (y <- x@c1) (1: y <- x) (b) (c),              ///
                              cov(d) cov(e) var(f) group(mygrp)
```

The expansion of (y <- x@c1) (1: y <- x) reads

```
(1: y <- x@c1)  (2: y <- x@c1)  (3: y <- x@c1)  (1: y <- x)
```

and you might think that 1: y <- x would replace 1: y <- x@c1. Information, however, is combined, and even though precedence is given to information appearing later, silence does not count as information. Thus the expanded and reduced specification reads the same as if 1: y <- x was never specified:

```
(1: y <- x@c1)  (2: y <- x@c1)  (3: y <- x@c1)
```

7. Items 1–6, stated in terms of *paths*, apply equally to what is typed inside the means(), variance(), and covariance() options. For instance, if you typed

```
. sem (a) (b) (c), var(e.y@c1) group(mygrp)
```

then you are constraining the variance to be equal across all three groups.

If you wanted to constrain the variance to be equal in groups 2 and 3, you could type

```
. sem (a) (b) (c), var(e.y) var(2: e.y@c1) var(3: e.y@c1), group(mygrp)
```

You could omit typing var(e.y) because it is implied. Alternatively, you could type

```
. sem (a) (b) (c), var(e.y@c1) var(1: e.y@c2) group(mygrp)
```

You could not type

```
. sem (a) (b) (c), var(e.y@c1) var(1: e.y) group(mygrp)
```

because silence does not count as information when specifications are combined.

Similarly, if you typed

```
. sem (a) (b) (c), cov(e.y1*e.y2@c1) group(mygrp)
```

then you are constraining the covariance to be equal across all groups. If you wanted to constrain the covariance to be equal in groups 2 and 3, you could type

```
. sem (a) (b) (c), cov(e.y1*e.y2)                                ///
                              cov(2: e.y1*e.y2@c1) cov(3: e.y1*e.y2@c1) ///
                              group(mygrp)
```

You could not omit cov(e.y1*e.y2) because it is not assumed. By default, error variables are assumed to be uncorrelated. Omitting the option would constrain the covariance to be 0 in group 1, and to be equal in groups 2 and 3.

Alternatively, you could type

```
. sem (a) (b) (c), cov(e.y1*e.y2@c1)                    ///
                   cov(1: e.y1*e.y2@c2)                 ///
                   group(mygrp)
```

8. In the examples above, we have referred to the groups using their numeric values, 1, 2, and 3. Had the values been 5, 8, and 10, then we would have used those values.

   If the group variable mygrp has a value label, you can use the label to refer to the group. For instance, imagine mygrp is labeled as follows:

```
. label define grpvals 1 Male  2 Female  3 "Unknown sex"
. label values mygrp grpvals
```

   We could type

```
. sem (y <- x) (Female: y <- x@c1) (Unknown sex: y <- x@c1) ..., ...
```

   or we could type

```
. sem (y <- x) (2: y <- x@c1) (3: y <- x@c1) ..., ...
```

Also see

[SEM] **sem** — Structural equation model estimation command

[SEM] **intro 2** — Learning the language: Path diagrams and command language

[SEM] **intro 5** — Comparing groups

Title

sem postestimation — Postestimation tools for sem

Syntax

The following are the postestimation commands that you can use after estimation by `sem`:

Command	Description
`sem`	without arguments, redisplays results
`sem, coeflegend`	display `_b[]` notation
`estat framework`	display results in modeling framework (matrix form)
`estat gof`	overall goodness of fit
`estat ggof`	group-level goodness of fit
`estat eqgof`	equation-level goodness of fit
`estat residuals`	matrices of residuals
`estat ic`	AIC and BIC statistics
`estat mindices`	modification indices (score tests)
`estat scoretests`	score tests
`estat ginvariant`	test of invariance of parameters across groups
`estat eqtest`	equation-level Wald tests
`lrtest`	likelihood-ratio tests
`test`	Wald tests
`lincom`	linear combination of parameters
`nlcom`	nonlinear combination of parameters
`testnl`	Wald tests of nonlinear hypotheses
`estat stdize:`	test standardized parameters
`estat teffects`	decomposition of effects
`estat stable`	assess stability of nonrecursive systems
`estat summarize`	display estimation-sample summary statistics
`estat vce`	display variance–covariance matrix of estimates
`predict`	factor scores, predicted values, etc.
`estimates`	cataloging estimation results

Description

For a summary of postestimation features, see [SEM] **intro 6**.

Remarks

estat ic, estat summarize, and estat vce are the standard estat commands available after all estimation commands; see [R] **estat**. Also see [SEM] **estat summarize** and [SEM] **estat gof**.

estimates is another feature available after all estimation commands that allows the storage and manipulation of estimation results both in memory and on disk; see [R] **estimates**.

Also see

[SEM] **sem reporting options** — Options affecting reporting of results

[SEM] **estat framework** — Display estimation results in modeling framework

[SEM] **estat gof** — Goodness-of-fit statistics

[SEM] **estat ggof** — Group-level goodness-of-fit statistics

[SEM] **estat eqgof** — Equation-level goodness-of-fit statistics

[SEM] **estat residuals** — Display mean and covariance residuals

[SEM] **estat mindices** — Modification indices

[SEM] **estat scoretests** — Score tests

[SEM] **estat ginvariant** — Tests for invariance of parameters across groups

[SEM] **estat eqtest** — Equation-level test that all coefficients are zero

[SEM] **test** — Wald test of linear hypotheses

[SEM] **lrtest** — Likelihood-ratio test of linear hypothesis

[SEM] **estat stdize** — Test standardized parameters

[SEM] **estat teffects** — Decomposition of effects into total, direct, and indirect

[SEM] **estat stable** — Check stability of nonrecursive system

[SEM] **estat summarize** — Report summary statistics for estimation sample

[R] **estat** — Postestimation statistics

[SEM] **lincom** — Linear combinations of parameters

[SEM] **nlcom** — Nonlinear combinations of parameters

[SEM] **predict** — Factor scores, linear predictions, etc.

[R] **estimates** — Save and manipulate estimation results

Title

sem reporting options — Options affecting reporting of results

Syntax

sem *paths* ..., ... *reporting_options*

sem, *reporting_options*

reporting_options	Description
level(#)	set confidence level; default is level(95)
standardized	display standardized coefficients and values
coeflegend	display coefficient legend
nocnsreport	do not display constraints
nodescribe	do not display variable classification table
noheader	do not display header above parameter table
nofootnote	do not display footnotes below parameter table
notable	do not display parameter table
nolabel	display group values rather than value labels
wrap(#)	allow long group label to wrap the first # lines
showginvariant	report all estimated parameters

Description

These options control how sem displays estimation results.

Options

level(#); see [R] **estimation options**.

standardized displays standardized values, which is to say, "beta" values for coefficients, correlations for covariances, and 1s for variances. Standardized values are obtained using model-fitted variances (Bollen 1989, 124–125). We recommend caution in the interpretation of standardized values, especially with multiple groups.

coeflegend displays the legend that reveals how to specify estimated coefficients in _b[] notation, which you are sometimes required to type in specifying postestimation commands.

nocnsreport suppresses the display of the constraints. Fixed-to-zero constraints that are automatically set by sem are not shown in the report to keep the output manageable.

nodescribe suppresses display of the variable classification table.

noheader suppresses the header above the parameter table, the display that reports the final log-likelihood value, number of observations, etc.

nofootnote suppresses the footnotes displayed below the parameter table.

notable suppresses the parameter table.

`nolabel` displays group values rather than value labels.

`wrap(#)` allows long group labels to wrap the first # lines in the parameter table. The default is `wrap(0)`, which means that long group labels will be abbreviated to fit on a single line.

`showginvariant` specifies that each estimated parameter be reported in the parameter table. The default is to report each invariant parameter only once.

Remarks

Any of the above options may be specified when you fit the model or when you redisplay results, which you do by specifying nothing but options after the `sem` command:

```
. sem (...) (...), ...
(original output displayed)
. sem
(output redisplayed)
. sem, standardized
(standardized output displayed)
. sem, coeftable
(coefficient table displayed)
. sem
(output redisplayed)
```

Also see

[SEM] **sem** — Structural equation model estimation command

[SEM] **example 8** — Testing that coefficients are equal, and constraining them

[SEM] **example 16** — Correlation

Title

sem ssd options — Options for use with summary statistics data

Syntax

`sem` *paths* ..., ... *ssd_options*

ssd_options	Description
`select()`	alternative to `if` *exp* for SSD
`forcecorrelations`	allow groups and pooling of SSD correlations

Description

Data are sometimes available in summary statistics form only. These summary statistics include means, standard deviations or variances, and correlations or covariances. These summary statistics can be used by `sem` in place of the underlying raw data.

Options

`select()` is an alternative to `if` *exp* when you are using summary statistics data (SSD). Where you might usually type

```
. sem ... if agegrp==1 | agegrp==3 | agegrp==5, ...
```

with SSD in memory, you type

```
. sem ..., ... select(1 3 5)
```

See [SEM] **sem option select()** and [SEM] **intro 10**.

`forcecorrelations` tells `sem` that it may make calculations that would usually be considered suspicious using SSD that contain only a subset of means, variances (standard deviations), and covariances (correlations). Do not specify this option unless you appreciate the statistical issues that we are about to discuss. There are two cases where `forcecorrelations` is relevant.

In the first case, `sem` is unwilling to produce `group()` estimates if one or more (usually all) of the groups have correlations only defined. You can override that by specifying `forcecorrelations`, and `sem` will assume unit variances for the group or groups that have correlations only. Doing this is suspect unless you make `ginvariant()` all parameters that are dependent on covariances or unless you truly know that the variances are indeed 1.

In the second case, `sem` is unwilling to pool across groups unless you have provided means and covariances (or means and correlations and standard deviations or variances). Without that information, should the need for pooling arise, `sem` issues an error message. The `forcecorrelations` option specifies that `sem` ignore its rule and pool correlation matrices, treating correlations as if they were covariances when variances are not defined and treating means as if they were 0 when means are not defined. The only justification for making the calculation in this way is that variances truly are 1 and means truly are 0.

Understand that there is nothing wrong with using pure correlation data, or covariance data without the means, so long as you fit models for individual groups. Doing anything across groups basically requires that sem have the covariance and mean information.

Remarks

See [SEM] **intro 10**.

Also see

[SEM] **sem** — Structural equation model estimation command

[SEM] **intro 10** — Fitting models using summary statistics data

[SEM] **ssd** — Making summary statistics data

Title

sem syntax options — Options affecting interpretation of syntax

Syntax

sem *paths* ..., ... *syntax_options*

syntax_options	Description
latent(*names*)	explicitly specify latent variable names
nocapslatent	do not treat capitalized *Names* as latent

where *names* is a space-separated list of the names of the latent variables.

Description

These options affect some minor issues of how sem interprets what you type.

Options

latent(*names*) specifies that *names* is the full set of names of the latent variables. sem ordinarily assumes that latent variables are the variables that have the first letter capitalized and observed variables have the first letter lowercased; see [SEM] **sem path notation**. When you specify latent(*names*), sem treats the listed variables as the latent variables and all other variables, regardless of capitalization, as observed. latent() implies nocapslatent.

nocapslatent specifies that having the first letter capitalized does not designate a latent variable. This option can be used when fitting models with observed variables only where some observed variables in the dataset have the first letter capitalized.

Remarks

We recommend using sem's default naming convention. If your dataset contains variables with the first letter capitalized, it is easy to convert the variables to have lowercase names by typing

```
. rename *, lower
```

See [D] **rename group**.

Also see

[SEM] **sem** — Structural equation model estimation command

[SEM] **sem path notation** — Command syntax for path diagrams

Title

ssd — Making summary statistics data

Syntax

To enter summary statistics data, the commands are

```
ssd init varlist

ssd set [#] observations #

ssd set [#] means vector

ssd set [#] {variances | sd} vector

ssd set [#] {covariances | correlations} matrix


ssd addgroup varname                     (to add the second group)

ssd addgroup                             (to add subsequent groups)

ssd unaddgroup #                         (to remove last group)


ssd status [#]                           (to review status)
```

To build summary statistics data from raw data, the command is

```
ssd build varlist [, group(varname) clear]
```

To review the contents of summary statistics data, the commands are

```
ssd status [#]

ssd describe

ssd list [#]
```

In an emergency (ssd will tell you when), you may use

```
ssd repair
```

In the above,

A *vector* can be any of the following:

1. A space-separated list of numbers, for example,

```
. ssd set means 1 2 3
```

2. `(stata)` *matname*
   where *matname* is the name of a Stata $1 \times k$ or $k \times 1$ matrix, for example,

```
. ssd set variances (stata) mymeans
```

3. `(mata)` *matname*
   where *matname* is the name of a Mata $1 \times k$ or $k \times 1$ matrix, for example,

```
. ssd set sd (mata) mymeans
```

A *matrix* can be any of the following:

1. A space-separated list of numbers corresponding to the rows of the matrix, with backslashes (\) between rows. The numbers are either the lower triangle and diagonal or the diagonal and upper triangle of the matrix, for example,

```
. ssd set correlations 1 \ .2 1 \ .3 .5 1
```

   or

```
. ssd set correlations 1 .2 .3 \ 1 .5 \ 1
```

2. `(ltd)` # # ...
   which is to say, a space-separated list of numbers corresponding to the lower triangle and diagonal of the matrix, without backslashes between rows, for example,

```
. ssd set correlations (ltd) 1  .2 1  .3 .5 1
```

3. `(dut)` # # ...
   which is to say, a space-separated list of numbers of the diagonal and upper triangle of the matrix, without backslashes between rows, for example,

```
. ssd set correlations (dut) 1 .2 .3   1 .5   1
```

4. `(stata)` *matname*
   where *matname* is the name of a Stata $k \times k$ symmetric matrix, for example,

```
. ssd set correlations (stata) mymat
```

5. `(mata)` *matname*
   where *matname* is the name of a Mata $k \times k$ symmetric matrix, for example,

```
. ssd set correlations (mata) mymat
```

Description

`ssd` allows you (1) to enter summary statistics data to fit structural equation models and (2) to create summary statistics data from original, raw data to publish or to send to others (and thus preserve participant confidentiality).

Options

`group(`*varname*`)` is for use with `ssd build`. It specifies that separate groups of summary statistics be produced for each value of *varname*.

`clear`, for use with `ssd build`, specifies that it is okay to replace the data in memory with summary statistics data even if the original dataset has not been saved since it was last changed.

Remarks

See

[SEM] **intro 10** Fitting models using summary statistics data

for an introduction, and see

[SEM] **example 2** Creating datasets from published covariances
[SEM] **example 19** Creating multiple-group summary statistics data
[SEM] **example 25** Creating summary statistics data from raw data

A summary statistics dataset is different from a regular, raw Stata dataset. Be careful not to use standard Stata data-manipulation commands with summary statistics data in memory. The commands include

```
generate
replace
merge
append
drop
set obs
```

to mention a few. You may, however, use `rename` to change the names of the variables.

The other data-manipulation commands can damage your summary statistics dataset. If you make a mistake and do use one of these commands, do not attempt to repair the data yourself. Let `ssd` repair your data by typing

```
. ssd repair
```

`ssd` is usually successful as long as variables or observations have not been dropped.

Every time you use `ssd`, even for something as trivial as describing or listing the data, `ssd` verifies that the data are not corrupted. If `ssd` finds that they are, it suggests you type `ssd repair`:

```
. ssd describe
SSD corrupt
    The summary statistics data should [ssd
    describes the problem]. The data may be fixable;
    type ssd repair.
. ssd repair
  (data repaired)
. ssd describe
(usual output appears)
```

In critical applications, we also recommend you digitally sign your summary statistics dataset:

```
. datasignature set
5:5(65336):3718404259:2275399871        (data signature set)
```

Then at any future time, you can verify the data are still just as they should be:

```
. datasignature confirm
(data unchanged since 30jun2011 15:32)
```

The data signature is a function of the variable names. If you rename a variable—something that is allowed—then the data signature will change:

```
. rename varname newname
. datasignature confirm
(data have changed since 30jun2011 15:32)
r(9);
```

In that case, you can re-sign the data:

```
. datasignature set, reset
5:5(71728):3718404259:2275399871        (data signature set)
```

Before re-signing, however, if you want to convince yourself that the data are unchanged except for the variable name, type `datasignature report`. It is the part in parentheses of the signature that has to do with the variable names. `datasignature report` will tell you what the new signature would be and you can verify that the other components of the signature match.

See [D] **datasignature**.

Saved results

`ssd describe` saves the following in `r()`:

Scalars

`r(N)`	number of observations (total across groups)
`r(k)`	number of variables (excluding group variable)
`r(G)`	number of groups
`r(complete)`	1 or 0; 1 if complete
`r(complete_means)`	1 or 0; 1 if complete means
`r(complete_covariances)`	1 or 0; 1 if complete covariances

Macros

`r(v#)`	variable names (excluding group variable)
`r(groupvar)`	name of group variable (if there is one)

Also see

[SEM] **intro 10** — Fitting models using summary statistics data

[D] **datasignature** — Determine whether data have changed

[SEM] **example 2** — Creating a dataset from published covariances

[SEM] **example 19** — Creating multiple-group summary statistics data

[SEM] **example 25** — Creating summary statistics data from raw data

Title

test — Wald test of linear hypotheses

Syntax

`sem ..., ...` (fit constrained or unconstrained model)

`test` *coeflist*

`test` *exp* `=` *exp* `=` [`=` ...]

`test` `[`*eqno*`]` [`:` *coeflist*]

`test` `[`*eqno* `=` *eqno* [`=` ...]`]` [`:` *coeflist*]

`test` `(`*spec*`)` [`(`*spec*`)` ...] [`,` *test_options*]

Menu

Statistics > Structural equation modeling (SEM) > Testing and CIs > Wald tests of linear hypotheses

Description

`test` performs the Wald test of the hypothesis or hypotheses that you specify.

`test` is a standard postestimation command and works after `sem` just as it does after any other estimation command except that you must use the `_b[]` coefficient notation; you cannot refer to variables using shortcuts to obtain coefficients on variables.

See [R] **test**. Also documented there is `testparm`. That command is not relevant after estimation by `sem` because its syntax hinges on use of shortcuts for referring to coefficients.

Options

See *Options for test* in [R] **test**.

Remarks

See [SEM] **example 8** and [SEM] **example 16**.

`test` works in the metric of SEM, which is to say, path coefficients, variances, and covariances. If you want to frame your tests in terms of standardized coefficients and correlations, prefix `test` with `estat stdize:`; see [SEM] **estat stdize**.

Saved results

See *Saved results* in [R] **test**.

Also see

[SEM] **example 8** — Testing that coefficients are equal, and constraining them

[SEM] **example 16** — Correlation

[R] **test** — Test linear hypotheses after estimation

[SEM] **estat stdize** — Test standardized parameters

[SEM] **estat eqtest** — Equation-level test that all coefficients are zero

[SEM] **lrtest** — Likelihood-ratio test of linear hypothesis

[SEM] **lincom** — Linear combinations of parameters

Title

testnl — Wald test of nonlinear hypotheses

Syntax

```
sem ..., ...                                   (fit constrained or unconstrained model)

testnl exp = exp [= ...] [, options]

testnl (exp = exp [= ...]) [(exp = exp [= ...]) ...] [, options]
```

Menu

Statistics > Structural equation modeling (SEM) > Testing and CIs > Wald tests of nonlinear hypotheses

Description

`testnl` performs the Wald test of the nonlinear hypothesis or hypotheses that you specify.

`testnl` is a standard postestimation command and works after `sem` just as it does after any other estimation command except that you must use the `_b[]` coefficient notation; you cannot refer to variables using shortcuts to obtain coefficients on variables.

Options

See *Options* in [R] **testnl**.

Remarks

`testnl` works in the metric of SEM, which is to say, path coefficients, variances, and covariances. If you want to frame your tests in terms of standardized coefficients and correlations, prefix `testnl` with `estat stdize:`; see [SEM] **estat stdize**.

❑ Technical note

`estat stdize:` is unnecessary because, using `testnl`, everywhere you wanted a standardized coefficient or correlation, you could just type the formula. If you did that, you would get the same answer but for numerical precision. In this case, the answer produced with the `estat stdize:` prefix will be a little more accurate because `estat stdize:` is able to substitute an analytic derivative in one part of the calculation where `testnl`, doing the whole thing itself, would be forced to use a numeric derivative.

❑

Saved results

See *Saved results* in [R] **testnl**.

Also see

[R] **testnl** — Test nonlinear hypotheses after estimation

[SEM] **test** — Wald test of linear hypotheses

[SEM] **lrtest** — Likelihood-ratio test of linear hypothesis

[SEM] **estat stdize** — Test standardized parameters

[SEM] **estat eqtest** — Equation-level test that all coefficients are zero

[SEM] **nlcom** — Nonlinear combinations of parameters

Glossary

ADF, method(adf). ADF stands for asymptotic distribution free and is a method used to obtain fitted parameters. ADF is used by `sem` when option `method(adf)` is specified. Other available methods are ML, QML, and MLMV.

anchoring, anchor variable. A variable is said to be the anchor of a latent variable if the path coefficient between the latent variable and the anchor variable is constrained to be 1. The `sem` software uses anchoring as a way of normalizing latent variables and thus identifying the model.

baseline model. A baseline model is a covariance model—a model of fitted means and covariances of observed variables without any other paths—with most of the covariances constrained to 0. That is, a baseline model is a model of fitted means and variances but typically not all the covariances. Also see *saturated model.*

Bentler–Weeks formulation. The Bentler and Weeks (1980) formulation of SEM places the results in a series of matrices organized around how results are calculated. See [SEM] **estat framework**.

bootstrap, vce(bootstrap). The bootstrap is a replication method for obtaining variance estimates. Consider an estimation method E for estimating θ. Let $\widehat{\theta}$ be the result of applying E to dataset D containing N observations. The bootstrap is a way of obtaining variance estimates for $\widehat{\theta}$ from repeated estimates $\widehat{\theta}_1$, $\widehat{\theta}_2$, ..., where each $\widehat{\theta}_i$ is the result of applying E to a dataset of size N drawn with replacement from D. See [SEM] **sem option method( )** and [R] **bootstrap**.

CI. CI is an abbreviation for confidence interval.

clustered, vce(cluster clustvar). Clustered is the name we use for the generalized Huber/White/sandwich estimator of the VCE, which is the `robust` technique generalized to relax the assumption that errors are independent across observations to be that they are independent across clusters of observations. Within cluster, errors may be correlated.

Clustered standard errors are reported when `sem` option `vce(cluster` *clustvar*`)` is specified. The other available techniques are OIM, EIM, OPG, robust, bootstrap, and jackknife.

CFA, CFA models. CFA stands for confirmatory factor analysis. It is a way of analyzing measurement models. CFA models is a synonym for measurement models.

coefficient of determination. The coefficient of determination is the fraction (or percentage) of variation (variance) explained by an equation of a model. The coefficient of determination is thus like R^2 in linear regression.

command language. Stata's `sem` command provides a way to specify structural equation models. The alternative is to use `sem`'s GUI to draw path diagrams; see [SEM] **intro 2** and [SEM] **GUI**.

constraints. See *parameter constraints.*

correlated uniqueness model. A correlated uniqueness model is a kind of measurement model in which the errors of the measurements has a structured correlation. See [SEM] **intro 4**.

curved path. See *path.*

degree-of-freedom adjustment. In estimates of variances and covariances, a finite-population degree-of-freedom adjustment is sometimes applied to make the estimates unbiased.

Let's write an estimated variance as $\widehat{\sigma}_{ii}$ and write the "standard" formula for the variance as $\widehat{\sigma}_{ii} = S_{ii}/N$. If $\widehat{\sigma}_{ii}$ is the variance of observable variable x_i, it can readily be proven that S_{ii}/N is a biased estimate of the variances in samples of size N and that $S_{ii}/(N-1)$ is an unbiased estimate. It is usual to calculate variances using $S_{ii}/(N-1)$, which is to say, the "standard" formula has a multiplicative degree-of-freedom adjustment of $N/(N-1)$ applied to it.

If $\widehat{\sigma}_{ii}$ is the variance of estimated parameter β_i, a similar finite-population degree-of-freedom adjustment can sometimes be derived that will make the estimate unbiased. For instance, if β_i is a coefficient from a linear regression, an unbiased estimate of the variance of regression coefficient β_i is $S_{ii}/(N-p-1)$, where p is the total number of regression coefficients estimated excluding the intercept. In other cases, no such adjustment can be derived. Such estimators have no derivable finite-sample properties and one is left only with the assurances provided by its provable asymptotic properties. In such cases, the variance of coefficient β_i is calculated as S_{ii}/N, which can be derived on theoretical grounds. SEM is an example of such an estimator.

SEM is a remarkably flexible estimator and can reproduce results that can sometimes be obtained by other estimators. SEM might produce asymptotically equivalent results, or it might produce identical results depending on the estimator. Linear regression is an example in which `sem` produces identical results. The reported standard errors, however, will not look identical because the linear regression estimates have the finite-population degree-of-freedom adjustment applied to them, and the SEM estimates do not. To see the equivalence, you must undo the adjustment on the reported linear regression standard errors by multiplying them by $\sqrt{\{(N-p-1)/N\}}$.

direct, **indirect**, and **total effects**. Consider the following system of equations:

$$\begin{aligned} y_1 &= b_{10} + b_{11}y_2 + b_{12}x_1 + b_{13}x_3 + e_1 \\ y_2 &= b_{20} + b_{21}y_3 + b_{22}x_1 + b_{23}x_4 + e_2 \\ y_3 &= b_{30} + \qquad\quad\;\; b_{32}x_1 + b_{33}x_5 + e_3 \end{aligned}$$

The total effect of x_1 on y_1 is $b_{12} + b_{11}b_{22} + b_{11}b_{21}b_{32}$. It measures the full change in y_1 based on allowing x_1 to vary throughout the system.

The direct effect of x_1 on y_1 is b_{12}. It measures the change in y_1 caused by a change in x_1 holding other endogenous variables—namely, y_2 and y_3—constant.

The indirect effect of x_1 on y_1 is obtained by subtracting the total and direct effect and is thus $b_{11}b_{22} + b_{11}b_{21}b_{32}$.

EIM, **vce(eim)**. EIM stands for expected information matrix, defined as the inverse of the negative of the expected value of the matrix of second derivatives, usually of the log-likelihood function. The EIM is an estimate of the VCE. EIM standard errors are reported when `sem` option `vce(eim)` is specified. The other available techniques are OIM, OPG, robust, clustered, bootstrap, and jackknife.

estimation method. There are a variety of ways that one can solve for the parameters of a structural equation model. Different methods make different assumptions about the data-generation process, and so it is important that you choose a method appropriate for your model and data; see [SEM] **intro 3**.

error, **error variable**. The error is random disturbance e in a linear equation:

$$y = b_0 + b_1x_1 + b_2x_2 + \cdots + e$$

An error variable is an unobserved exogenous variable in path diagrams corresponding to e. Mathematically, error variables are just another example of latent exogenous variables, but in `sem`, error variables are considered to be in a class by themselves. All endogenous variables—observed and latent—have a corresponding error variable. Error variables automatically and inalterably have their path coefficients fixed to be 1. Error variables have a fixed naming convention in the software. If a variable is the error for (observed or latent) endogenous variable `y`, then the residual variable's name is `e.y`.

In `sem`, error variables are uncorrelated with each other unless explicitly indicated otherwise. That indication is made in path diagrams by drawing a curved path between the error variables and is indicated in command notation by including `cov(e.`*name1*`*e.`*name2*`)` among the options specified on the `sem` command.

endogenous variable. A variable, observed or latent, is endogenous (determined by the system) if any path points to it. Also see *exogenous variable*.

exogenous variable. A variable, observed or latent, is exogenous (determined outside the system) if paths only originate from it, or equivalently, no path points to it. Also see *endogenous variable*.

fictional data. Fictional data are data that have no basis in reality even though they might look real; they are data that are made up for use in examples.

first- and second-order latent variables. If a latent variable is measured by other latent variables only, the latent variable that does the measuring are called first-order latent variable, and the latent variable being measured is called the second-order latent variable.

GMM, generalized method of moments. GMM is a method used to obtain fitted parameters. In this documentation, GMM is referred to as ADF, which stands for asymptotic distribution free. Other available methods are ML, QML, ADF, and MLMV.

The SEM moment conditions are cast in terms of second moments, not the first moments used in many other applications associated with GMM.

goodness-of-fit statistic. A goodness-of-fit statistic is a value designed to measure how well the model reproduces some aspect of the data the model is intended to fit. SEM reproduces the first- and second-order moments of the data, with an emphasis on the second-order moments, and thus goodness-of-fit statistics appropriate for use after `sem` compare the predicted covariance matrix (and mean vector) with the matrix (and vector) observed in the data.

GUI. GUI stands for graphical user interface and in this manual stands for the component of the software that allows you to specify models by entering path diagrams. The alternative way to enter your model is by using `sem`'s command language. See [SEM] **intro 2** and [SEM] **GUI**.

identification. Identification refers to the conceptual constraints on parameters of a model that are required for the model's remaining parameters to have a unique solution. A model is said to be unidentified if these constraints are not supplied. These constraints are of two types: substantive constraints and normalization constraints.

Normalization constraints deal with the problem that one scale works as well as another for each latent variable in the model. One can think, for instance, of propensity to write software as being measured on a scale of 0 to 1, 1 to 100, or any other scale. The normalization constraints are the constraints necessary to choose one particular scale. The normalization constraints are provided automatically by the `sem` software by anchoring using unit loadings.

Substantive constraints are the constraints you specify about your model so that it has substantive content. Usually, these constraints are zero constraints implied by the paths omitted, but they can include explicit parameter constraints as well. It is easy to write a model that is not identified for substantive reasons; See [SEM] **intro 3**.

indicator variables, indicators. Synonym for *measurement variables*.

indirect effects. See *direct, indirect, and total effects*.

initial values. See *starting values*.

intercept. An intercept for the equation of endogenous variable y, observed or latent, is the path coefficient from `_cons` to y. `_cons` is Stata-speak for the built-in variable containing 1 in all observations. In SEM-speak, `_cons` is an observed exogenous variable.

jackknife, vce(jackknife). The jackknife is a replication method for obtaining variance estimates. Consider an estimation method E for estimating θ. Let $\widehat{\theta}$ be the result of applying E to dataset D containing N observations. The jackknife is a way of obtaining variance estimates for $\widehat{\theta}$ from repeated estimates $\widehat{\theta}_1, \widehat{\theta}_2, \ldots, \widehat{\theta}_N$, where each $\widehat{\theta}_i$ is the result of applying E to D with observation i removed. See [SEM] **sem option method( )** and [R] **jackknife**.

Lagrange multiplier tests. Synonym for *score tests*.

linear regression. Linear regression is a kind of structural equation model in which there is a single equation and all values are observed. See [SEM] **intro 4**.

latent growth model. A latent growth model is a kind of measurement model in which the observed values are collected over time and are allowed to follow a trend. See [SEM] **intro 4**.

latent variable. A variable is latent if it is not observed. A variable is latent if it is not in your dataset but you wish it were. You wish you had a variable recording the propensity to commit violent crime, or socioeconomic status, or happiness, or true ability, or even income accurately recorded. Latent variables are sometimes described as imagined variables.

In the software, latent variables are usually indicated by having at least their first letter capitalized.

Also see *observed variables* and see *first- and second-order latent variables*.

manifest variables. Synonym for *observed variables*.

measure, measurement, x a measurement of X, x measures X. See *measurement variables*.

measurement models, measurement component. A measurement model is a particular kind of model that deals with the problem of translating observed values to values suitable for modeling. Measurement models are often combined with structural models and then the measurement model part is referred to as the measurement component. See [SEM] **intro 4**.

measurement variables, measure, measurement, x a measurement of X, x measures X. Observed variable x is a measurement of latent variable X if there is a path connecting $x \leftarrow X$. Measurement variables are modeled by measurement models. Measurement variables are also called indicator variables.

method. Method is just an English word and should be read in context. Nonetheless, method is used here usually to refer to the method used to solve for the fitted parameters of a structural equation model. Those methods are ML, QML, MLMV, and ADF. Also see technique.

MIMIC. MIMIC stands for multiple indicators and multiple causes. It is a kind of structural model in which observed causes determine a latent variable, which in turn determines multiple indicators. See [SEM] **intro 4**.

ML, method(ml). ML stands for maximum likelihood. It is a method to obtain fitted parameters. ML is the default method used by `sem`. Other available methods are QML, MLMV, and ADF.

MLMV, method(mlmv). MLMV stands for maximum likelihood with missing values. It is an ML method used to obtain fitted parameters in the presence of missing values. MLMV is the method used by `sem` when the `method(mlmv)` option is specified. Other available methods are ML, QML, and ADF. Those methods omit from the calculation observations that contain missing values.

modification indices. Modification indices are score tests for adding paths where none appear. The paths can be for either coefficients or covariances.

moments (of a distribution). The moments of a distribution are the expected values of powers of a random variable or centralized (demeaned) powers of a random variable. The first moments are the expected or observed means, and the second moments are the expected or observed variances and covariances.

multiple correlation. The multiple correlation is the correlation between endogenous variable y and its linear prediction.

multivariate regression. Multivariate regression is a kind of structural model in which each member of a set of observed endogenous variables is a function of the same set of observed exogenous variables and a unique random disturbance term. The disturbances are correlated. Multivariate regression is a special case of *seemingly unrelated regression*.

nonrecursive (structural) model (system), recursive (structural) model (system). A structural model (system) is said to be nonrecursive if there are paths in both directions between one or more pairs of endogenous variables. A system is recursive if it is a system—it has endogenous variables that appear with paths from them—and it is not nonrecursive.

A nonrecursive model may be unstable. Consider, for instance,

$$y_1 = 2y_2 + 1x_1 + e_1$$
$$y_2 = 3y_1 - 2x_2 + e_2$$

This model is unstable. To see this, without loss of generality, treat $x_1 + e_1$ and $2x_2 + e_2$ as if they were both 0. Consider $y_1 = 1$ and $y_2 = 1$. Those values result in new values $y_1 = 2$ and $y_2 = 3$, and those result in new values $y_1 = 6$ and $y_2 = 6$, and those result in new values, ... Continue in this manner, and you reach infinity for both endogenous variables. In the jargon of the mathematics used to check for this property, the eigenvalues of the coefficient matrix lie outside the unit circle.

On the other hand, consider these values:

$$y_1 = 0.5y_2 + 1x_1 + e_1$$
$$y_2 = 1.0y_1 - 2x_2 + e_2$$

These results are stable in that the resulting values converge to $y_1 = 0$ and $y_2 = 0$. In the jargon of the mathematics used to check for this property, the eigenvalues of the coefficients matrix lie inside the unit circle.

Finally, consider the values

$$y_1 = 0.5y_2 + 1x_1 + e_1$$
$$y_2 = 2.0y_1 - 2x_2 + e_2$$

Start with $y_1 = 1$ and $y_2 = 1$ and that yields new values $y_1 = 0.5$ and $y_2 = 2$ and that yields new values $y_1 = 1$ and $y_2 = 1$, and that yields $y_1 = 0.5$ and $y_2 = 2$, and it will oscillate forever. In the jargon of the mathematics used to check for this property, the eigenvalues of the coefficients matrix lie on the unit circle. These coefficients are also considered to be unstable.

normalization constraints. See *identification*.

normalized residuals. See *standardized residuals*.

observed variables. A variable is observed if it is a variable in your dataset. In this documentation, we often refer to observed variables using `x1`, `x2`, ..., `y1`, `y2`, and so on, but in reality observed variables have names such as `mpg`, `weight`, `testscore`, etc.

In the software, observed variables are usually indicated by having names that are all lowercase.

Also see *latent variable*.

OIM, vce(oim). OIM stands for observed information matrix, defined as the inverse of the negative of the matrix of second derivatives, usually of the log likelihood function. The OIM is an estimate of the VCE. OIM is the default VCE that `sem` reports. The other available techniques are EIM, OPG, robust, clustered, bootstrap, and jackknife.

OPG, vce(opg). OPG stands for outer product of the gradients, defined as the cross product of the observation-level first derivatives, usually of the log likelihood function. The OPG is an estimate of the VCE. The other available techniques are OIM, EIM, robust, clustered, bootstrap, and jackknife.

p-value. P-value is another term for the reported significance level associated with a test. Small p-values indicate rejection of the null hypothesis.

parameter constraints. Parameter constraints are restrictions placed on the parameters of the model. These constraints are typically in the form of zero constraints and equality constraints. A zero constraint is implied, for instance, when no path is drawn connecting x with y. An equality constraint is specified when one forces one path coefficient to be equal to another, or one covariance to be equal to another.

Also see *identification*.

parameters. The parameters are the to-be-estimated coefficients of a model. These include all path coefficients, means, variances, and covariances. In mathematical notation, the theoretical parameters are often written as $\boldsymbol{\theta} = (\boldsymbol{\alpha}, \boldsymbol{\beta}, \boldsymbol{\mu}, \boldsymbol{\Sigma})$, where $\boldsymbol{\alpha}$ is the vector of intercepts, $\boldsymbol{\beta}$ is the vector of path coefficients, $\boldsymbol{\mu}$ is the vector of means, and $\boldsymbol{\Sigma}$ is the matrix of variances and covariances. The resulting parameters estimates are written as $\widehat{\boldsymbol{\theta}}$.

path. A path, typically shown as an arrow drawn from one variable to another, states that the first variable determines the second variable, at least partially. If $x \rightarrow y$, or equivalently $y \leftarrow x$, then $y_j = \alpha + \cdots + \beta x_j + \cdots + e.y_j$, where β is said to be the $x \rightarrow y$ path coefficient. The ellipses are included to account for paths to y from other variables. α is said to be the intercept and is automatically added when the first path to y is specified.

A curved path is a curved line connecting two variables, and it specifies that the two variables are allowed to be correlated. If there is no curved path between variables, the variables are usually assumed to be uncorrelated. We say usually because correlation is assumed among observed exogenous variables and, in the command language, assumed among latent exogenous variables, and if some of the correlations are not desired, they must be suppressed. Many authors refer to covariances rather than correlations. Strictly speaking, the curved path denotes a nonzero covariance. A correlation is often called a standardized covariance.

A curved path can connect a variable to itself and in that case, indicates a variance. In path diagrams in this manual, we typically do not show such variance paths even though variances are assumed.

path coefficient. The path coefficient is associated with a path; see *path*. Also see *intercept*.

path diagram. A path diagram is a graphical representation that shows the relationships among a set of variables using *paths*. See [SEM] **intro 2** for a description of path diagrams.

path notation. Path notation is a syntax defined by the authors of Stata's `sem` command for entering path diagrams in a command language. Models to be fit may be specified in path notation or they may be drawn using path diagrams into `sem`'s GUI.

QML, method(ml) vce(robust). QML stands for quasimaximum likelihood. It is a method used to obtain fitted parameters, and a technique used to obtain the corresponding VCE. QML is used by `sem` when options `method(ml)` and `vce(robust)` are specified. Other available methods are ML, MLMV, and ADF. Other available techniques are OIM, EIM, OPG, clustered, bootstrap, and jackknife.

regression. A regression is a model in which an endogenous variable is written as a function of other variables, parameters to be estimated, and a random disturbance.

reliability. Reliability is the proportion of the variance of a variable not due to measurement error. A variable without measure error has reliability 1.

residual. In this manual, we reserve the word residual for the difference between the observed and fitted moments of an SEM model. We use the word error for the disturbance associated with a linear equation; see *error*. Also see *standardized residuals*.

robust, vce(robust). Robust is the name we use here for the Huber/White/sandwich estimator of the VCE. This technique requires fewer assumptions than most other techniques. In particular, it merely assumes that the errors are independently distributed across observations and thus allows the errors to be heteroskedastic. Robust standard errors are reported when the `sem` option `vce(robust)` is specified. The other available techniques are OIM, EIM, OPG, clustered, bootstrap, and jackknife.

saturated model. A saturated model is a full covariance model—a model of fitted means and covariances of observed variables without any restrictions on the values. Also see *baseline model*.

score tests, Lagrange multiplier tests. A score test is a test based on first derivatives of a likelihood function. Score tests are especially convenient for testing whether constraints on parameters should be relaxed or parameters should be added to a model. Also see *Wald tests*.

scores. Scores has two unrelated meanings. First, scores are the observation-by-observation first-derivatives of the (quasi) log-likelihood function. When we use the word scores, this is what we mean. Second, in the factor-analysis literature, scores (usually in the context of factor scores) refers to the expected value of a latent variable conditional on all the observed variables. We refer to this simply as the predicted value of the latent variable.

second-order latent variable. See *first- and second-order latent variables*.

seemingly unrelated regression. Seemingly unrelated regression is a kind of structural model in which each member of a set of observed endogenous variables is a function of a set of observed exogenous variables and a unique random disturbance term. The disturbances are correlated and the sets of exogenous variables may overlap. If the sets of exogenous variables are identical, this is referred to as multivariate regression.

SEM. SEM stands for structural equation modeling and for structural equation model. We use SEM in capital letters when writing about theoretical or conceptual issues as opposed to issues of the particular implementation of SEM in Stata with the `sem` command.

sem. `sem` is the Stata command that fits structural equation models.

SSD, ssd. SSD stands for summary statistics data. Data are sometimes available only in summary statistics form, as (1) means and covariances, (2) means, standard deviations or variances, and correlations, (3) covariances, (4) standard deviations or variances and correlations, or (5) correlations. SEM can be used to fit models using such data in place of the underlying raw data. The `ssd` command creates datasets containing summary statistics.

standardized coefficient. In a linear equation $y = \dots\ bx + \dots$, the standardized coefficient β is $(\widehat{\sigma}_y/\widehat{\sigma}_x)b$. Standardized coefficients are scaled to units of standard-deviation change in y for a standard-deviation change in x.

standardized covariance. A standardized covariance between y and x is equal to the correlation of y and x, which is to say, it is equal to $\sigma_{xy}/\sigma_x\sigma_y$. The covariance is equal to the correlation when variables are standardized to have variance 1.

standardized residuals, normalized residuals. Standardized residuals are residuals adjusted so that they follow a standard normal distribution. The difficulty is that the adjustment is not always possible. Normalized residuals are residuals adjusted according to a different formula that roughly follow a standard normal distribution. Normalized residuals can always be calculated.

starting values. The estimation methods provided by `sem` are iterative. The starting values are values for each of the parameters to be estimated that are used to initialize the estimation process. The `sem` software provides starting values automatically, but in some cases, these are not good enough and you must (1) diagnose the problem and (2) provide better starting values. See [SEM] **intro 3** and see *How to solve convergence problems* in [SEM] **sem**.

structural equation model. Different authors use the term structural equation model in different ways, but all would agree that a structural equation model sometimes carries the connotation of being a structural model with a measurement component, which is to say, combined with a measurement model.

structural model. A structural model is a model in which the parameters are not merely a description but believed to be of a causal nature. Obviously, SEM can fit structural models and thus so can `sem`. Neither SEM nor `sem` are limited to fitting structural models, however.

Structural models often have multiple equations and dependencies between endogenous variables, although that is not a requirement.

See [SEM] **intro 4**. Also see *structural equation model.*

structured (correlation or covariance). See *unstructured and structured (correlation or covariance).*

substantive constraints. See *identification.*

summary statistics data. See SSD.

technique. Technique is just an English word and should be read in context. Nonetheless, technique is usually used here to refer to the technique used to calculate the estimated VCE. Those techniques are OIM, EIM, OPG, robust, clustered, bootstrap, and jackknife.

Technique is also used to refer to the available techniques used with `ml`, Stata's optimizer and likelihood maximizer, to find the solution.

total effects. See *direct, indirect, and total effects.*

unstandardized coefficient. A coefficient that is not standardized. If `mpg` $= -0.006 \times$ `weight` $+$ 39.44028, then -0.006 is an unstandardized coefficient and, as a matter of fact, is measured in mpg-per-pound units.

unstructured and structured (correlation or covariance). A set of variables, typically error variables, is said to have an unstructured correlation or covariance if the covariance matrix has no particular pattern imposed by theory. If a pattern is imposed, the correlation or covariance is said to be structured.

VCE, variance–covariance matrix (of the estimator). The estimator is the formula used to solve for the fitted parameters, sometimes called the fitted coefficients. The VCE is the estimated variance–covariance matrix of the parameters. The diagonal elements of the VCE are the variances of the parameters or equivalent, the square root of those elements are the reported standard errors of the parameters.

Wald tests. A Wald test is a statistical test based on the estimated variance–covariance matrix of the parameters. Wald tests are especially convenient for testing possible constraints to be placed on the estimated parameters of a model. Also see *score tests*.

WLS, weighted least squares. Weighted least squares is a method used to obtain fitted parameters. In this documentation, WLS is referred to as ADF, which stands for asymptotic distribution free. Other available methods are ML, QML, and MLMV. ADF is, in fact, a specific kind of the more generic WLS.

References

Akaike, H. 1974. A new look at the statistical model identification. *IEEE Transactions on Automatic Control* 19: 716–723.

——. 1987. Factor analysis and AIC. *Psychometrika* 52: 317–332.

Bentler, P. M. 1990. Comparative fit indexes in structural models. *Psychological Bulletin* 107: 238–246.

Bentler, P. M., and E. H. Freeman. 1983. Tests for stability in linear structural equation systems. *Psychometrika* 48: 143–145.

Bentler, P. M., and T. Raykov. 2000. On measures of explained variance in nonrecursive structural equation models. *Journal of Applied Psychology* 85: 125–131.

Bentler, P. M., and D. G. Weeks. 1980. Linear structural equations with latent variables. *Psychometrika* 45: 289–308.

Bollen, K. A. 1989. *Structural Equations with Latent Variables*. New York: Wiley.

Bollen, K. A., and P. J. Curran. 2006. *Latent Curve Models: A Structural Equation Perspective*. Hoboken, NJ: Wiley.

Brown, T. A. 2006. *Confirmatory Factor Analysis for Applied Research*. New York: Guilford Press.

Browne, M. W., and R. Cudeck. 1993. Alternative ways of assessing model fit. Reprinted in *Testing Structural Equation Models*, ed. K. A. Bollen and J. S. Long, pp. 136–162. Newbury Park, CA: Sage.

Campbell, D. T., and D. W. Fiske. 1959. Convergent and discriminant validation by the multitrait-multimethod matrix. *Psychological Bulletin* 56: 81–105.

Duncan, O. D., A. O. Haller, and A. Portes. 1968. Peer influences on aspirations: A reinterpretation. *American Journal of Sociology* 74: 119–137.

Hancock, G. R., and R. O. Mueller, ed. 2006. *Structural Equation Modeling: A Second Course*. Charlotte, NC: Information Age Publishing.

Hausman, J. A. 1978. Specification tests in econometrics. *Econometrica* 46: 1251–1271.

Jöreskog, K. G., and D. Sörbom. 1986. *Lisrel VI: Analysis of linear structural relationships by the method of maximum likelihood*. Mooresville, IN: Scientific Software.

Kenny, D. A. 1979. *Correlation and Causality*. New York: Wiley.

Kline, R. B. 2005. *Principles and Practice of Structural Equation Modeling*. 2nd ed. New York: Guilford Press.

——. 2011. *Principles and Practice of Structural Equation Modeling*. 3rd ed. New York: Guilford Press.

Kluegel, J. R., R. Singleton, Jr., and C. E. Starnes. 1977. Subjective class identification: A multiple indicator approach. *American Sociological Review* 42: 599–611.

Marsh, H. W., and D. Hocevar. 1985. Application of confirmatory factor analysis to the study of self-concept: First- and higher order factor models and their invariance across groups. *Psychological Bulletin* 97: 562–582.

Raftery, A. E. 1993. Bayesian model selection in structural equation models. Reprinted in *Testing Structural Equation Models*, ed. K. A. Bollen and J. S. Long, pp. 163–180. Newbury Park, CA: Sage.

Schwarz, G. 1978. Estimating the dimension of a model. *Annals of Statistics* 6: 461–464.

Sobel, M. E. 1987. Direct and indirect effects in linear structural equation models. *Sociological Methods and Research* 16: 155–176.

Sörbom, D. 1989. Model modification. *Psychometrika* 54: 371–384.

Wheaton, B., B. Muthén, D. F. Alwin, and G. F. Summers. 1977. Assessing reliability and stability in panel models. In *Sociological Methodology 1977*, ed. D. R. Heise, 84–136. San Francisco: Jossey-Bass.

Williams, T. O., Jr., R. C. Eaves, and C. Cox. 2002. Confirmatory factor analysis of an instrumental designed to measure affective and cognitive arousal. *Educational and Psychological Measurement* 62: 264–283.

Wooldridge, J. M. 2010. *Econometric Analysis of Cross Section and Panel Data*. 2nd ed. Cambridge, MA: MIT Press.

Subject and author index

This is the subject and author index for the *Structural Equation Modeling Reference Manual.* Readers interested in topics other than structural equation modeling should see the combined subject index (and the combined author index) in the *Quick Reference and Index.*

A

B

C

D

E

F

G

H

I

J

K

L

M

N